Evolutionary Conservation Genetics

Evolutionary Conservation Genetics

Jacob Höglund

OXFORD
UNIVERSITY PRESS

Great Clarendon Street, Oxford OX2 6DP

Oxford University Press is a department of the University of Oxford.
It furthers the University's objective of excellence in research, scholarship,
and education by publishing worldwide in

Oxford New York

Auckland Cape Town Dar es Salaam Hong Kong Karachi
Kuala Lumpur Madrid Melbourne Mexico City Nairobi
New Delhi Shanghai Taipei Toronto

With offices in

Argentina Austria Brazil Chile Czech Republic France Greece
Guatemala Hungary Italy Japan Poland Portugal Singapore
South Korea Switzerland Thailand Turkey Ukraine Vietnam

Oxford is a registered trade mark of Oxford University Press
in the UK and in certain other countries

Published in the United States
by Oxford University Press Inc., New York

© Jacob Höglund 2009

The moral rights of the author have been asserted
Database right Oxford University Press (maker)

First published 2009

All rights reserved. No part of this publication may be reproduced,
stored in a retrieval system, or transmitted, in any form or by any means,
without the prior permission in writing of Oxford University Press,
or as expressly permitted by law, or under terms agreed with the appropriate
reprographics rights organization. Enquiries concerning reproduction
outside the scope of the above should be sent to the Rights Department,
Oxford University Press, at the address above

You must not circulate this book in any other binding or cover
and you must impose the same condition on any acquirer

British Library Cataloguing in Publication Data

Data available

Library of Congress Cataloging in Publication Data

Data available

Typeset by Newgen Imaging Systems (P) Ltd., Chennai, India
Printed in Great Britain
on acid-free paper by
the MPG Books Group

ISBN 978–0–19–921421–1 (Hbk.)   978–0–19–921422–8 (Pbk.)

10 9 8 7 6 5 4 3 2 1

Contents

Preface and acknowledgements   viii

1   The extinction vortex, is genetic variation related to extinction?   1
    1.1   Introduction   1
    1.2   The extinction vortex   2
    1.3   Evidence from wild populations of a link between low genetic diversity and extinction   5
    1.4   Experimental studies   14
    1.5   Conclusions   17

2   How to measure genetic variation   18
    2.1   Codominant neutral variation   18
       2.1.1   Percentage of polymorphic loci   19
       2.1.2   Alleles per locus/allelic richness   22
       2.1.3   Expected heterozygosity   22
       2.1.4   Observed heterozygosity   22
       2.1.5   Inbreeding coefficient   22
       2.1.6   Population differentiation   22
       2.1.7   Gene flow   23
    2.2   Dominant neutral markers   23
    2.3   Sequence variation   24
       2.3.1   Proportion of variable sites   25
       2.3.2   Nucleotide diversity   25
       2.3.3   Haplotype diversity   26
    2.4   Non-neutral markers and neutrality tests   26
    2.5   Quantitative additive genetic variation   27
    2.6   Conclusions   36

3   Inbreeding, geographic subdivision, and gene flow   37
    3.1   Inbreeding within populations   37
    3.2   Population structure   45

	3.3	Effective population size	47
	3.4	Examples of population structure in endangered species	50
	3.5	Inbreeding depression	51
	3.6	Heterozygosity–fitness correlations	55
	3.7	Rescue effects	58
	3.8	Conclusions	59
4	**Genetic diversity in changing environments**	**60**
	4.1	Fragmentation and natural and human-induced barriers to gene flow	60
	4.2	Landscape genetics	69
	4.3	Effects of bottlenecks and how to detect them	73
	4.4	Effects of population expansions and range shifts	75
	4.5	Invasive species	78
	4.6	Summary	80
5	**Genes under selection: *Mhc* and others**	**81**
	5.1	*Mhc* genes	82
		5.1.1 *Mhc* and conservation in mammals	84
		5.1.2 *Mhc* and conservation in birds	85
		5.1.3 *Mhc* and conservation in reptiles and amphibians	88
		5.1.4 *Mhc* and conservation in fish	91
		5.1.5 Summary: *Mhc* and immunogenetics in conservation	94
	5.2	Other candidate genes relevant for conservation	95
		5.2.1 Pigmentation genes: *mc1r*	95
		5.2.2 Photoperiodism: *Clock* and other genes	98
	5.3	Conclusions	100
6	**Local adaptation**	**102**
	6.1	Evidence of local adaptation	103
	6.2	Differentiation in quantitative traits, Q_{ST}	108
	6.3	Comparisons of F_{ST} and Q_{ST}	109
	6.4	Q_{ST} applied to conservation studies	114
	6.5	Conclusions	116
7	**Ecological genomics**	**119**
	7.1	WGS	120
	7.2	What to do with the data? Assembly and annotation	121
	7.3	What to do with the data? Evolutionary and ecological analyses	122
	7.4	Genomics in conservation	129
		7.4.1 SNP detection and genotyping	129

		7.4.2 QTL mapping of functionally important loci	131
		7.4.3 Differential gene expression	132
		7.4.4 Phylogenetics	133
	7.5	Genomic studies of non-model species	134
	7.6	Conclusions	138
8	**An evolutionary conservation biology**		**139**
	8.1	Human impact on evolutionary processes	140
	8.2	Evolutionary responses of harvesting	142
	8.3	Conserving evolutionary potential	143
	8.4	Conservation units	145
	8.5	Concluding remarks	148
References			151
Index			185

Preface and acknowledgements

I had great difficulty finding a title for this book. For long, the working title was *Genetic Variation and Extinction*. However, this title implies a causal and simple relationship between genetic variation and extinction. I do think that the study of genetic variation is extremely important for conservation biology but, as will become apparent while reading the text, I am not as sure that this relationship is as simple and straightforward as I thought when I began this voyage. I then started to think of alternatives and found two: *Evolutionary Conservation Biology* and *Conservation Biology and Evolution*. Of these, the first one has already been used for the volume edited by Ferriere *et al.* (2004) and I was not happy with the other one. This book is about conservation biology, so the first part is fine, but by using the word *Evolution* in the title I would have had to put more emphasis on the history of life on Earth and on how genetic diversity has evolved on the planet Earth. That is not a topic of this book and therefore I preferred to use the word *Evolutionary*, which implies that evolutionary theory and thinking in a more general sense are a large part of the book. One early morning and during the final stages of writing, I woke up and I decided that the title should be *Conservation and Evolutionary Biology*. However, conservation biology is more than what is covered by this book. What I have done in the following is an attempt to cover the evolutionary aspects of the genetic parts of conservation biology; there are no attempts to review the issues of, for example, habitat management, restoration projects, and the socioeconomic aspects conservation. The final decision on the title was therefore *Evolutionary Conservation Genetics*.

I am indebted to the many people who have helped and aided me while writing this book. My colleagues at the Evolutionary Biology Centre at Uppsala University are acknowledged for providing a world 'read in tooth and claw'. Dianna Steiner assisted in creating the reference list and in my understanding of the mysteries of software for handling references. She also compiled a summary on landscape genetics which was very helpful. Hans Höglund assisted in preparing all the figures in a suitable digital format and also assisted with the handling of references. Ian Sherman, Helen Eaton, and the rest of the staff at Oxford University Press provided much support and understanding during all stages of the writing. Martin Lascoux, Ulf Lagercrantz, Mikael Lönn, Tanja Strand, Björn Rogell, Robert Ekblom, Stefan Palm, Martin Carlsson, and (unknowingly) Scott Edwards read parts of the book or provided hints and tips. Gernot Segelbacher is gratefully acknowledged for not only reading and commenting on the whole manuscript but also for his friendship and for a most helpful visit in the crazy days in June 2008 when I was approaching yet another deadline. Finally I thank my family for their love and support.

Jacob Höglund
June 2008

1 *The extinction vortex, is genetic variation related to extinction?*

1.1 **Introduction**

Extinction is a fact. Ever since organic life first evolved on this planet, life forms have been changing. New species have arisen and old ones have gone extinct (Raup 1992). Speciation, the birth of new species, and extinction, the death of species, are as natural events in evolution as birth and death of individuals in demography. Seen over the entire history of organic life on Earth, biodiversity has generally increased. There has been a build up of life forms. However, five times in the evolutionary past of the planet have mass extinction events taken place. The so-called big five are periods when the rate of extinction of species has become vastly elevated and have outnumbered the level of new species forming (Raup 1994). It is now established that some of the elevated levels of mass extinction coincide with major celestial impacts on the Earth's surface and their climatic consequences, although some workers advocate more complex scenarios that include a number of factors that may explain mass extinction (Erwin 2006). Today we are witnessing a sixth major mass extinction event and this time celestial impact has nothing to do with it. It is beyond doubt that this event is caused by the activities of one of the species inhabiting the Earth: modern humans. I can think of no other scientific activity more important than trying to understand the causes and consequences of this contemporary mass extinction. This book is therefore concerned with a proposition put forward some years ago that extinction of species is somehow related to loss of genetic variation.

It has been suggested that genetic variation is crucial for the persistence of populations (Soulé 1980, 1986, 1987, Frankel and Soulé 1981, Gilpin and Soulé 1986). Two reasons have been given. In the short term, inbreeding and genetic drift leads to lower fitness of individuals and increased extinction risk of populations. In the long term, populations that lose genetic variation cannot evolve since evolution cannot proceed without genetic variation. In a world of rapid environmental change, any population that is unable to adapt to changing conditions will go extinct (Spielman *et al.* 2004).

After initial enthusiasm over this idea much scepticism has been raised. In 1988, Russell Lande wrote an influential paper (Lande 1988) in which he discussed the arguments for and against demographic versus genetic reasons for extinction of endangered populations: "Theory and empirical examples suggest that demography is usually of more immediate importance than population genetics in determining the minimum viable sizes of wild populations. The practical need in biological conservation for understanding the interaction of demographic and genetic factors in extinction may provide a focus for fundamental advances at the interface of ecology and evolution". He thus argued that demographic factors were more important than genetics in explaining why populations go extinct but that the interaction between demography and genetics should be a research focus. Unfortunately the paper has often been cited as an argument against genetic studies in conservation biology (e.g. Pimm 1991, Young 1991, Wilson 1992, Caro and Laurenson 1994, Caughley 1994, Holsinger et al. 1999, Elgar and Clode 2001). Recently, a perhaps more balanced view has emerged, in which both genetic and demographic factors are believed to be important in the study of endangered populations and species (Soulé and Mills 1998, Hedrick 2001, Oostermeijer et al. 2003). This chapter is a review of genetic studies and examples that suggest a link between genetic diversity and population persistence.

1.2 The extinction vortex

Theoretical considerations suggest that small—that is, endangered—populations are different from large ones in two important aspects. The level of inbreeding is increased and likewise the importance of genetic drift, the stochastic loss of alleles, in shaping a population's genetic architecture is increased. Both these processes ultimately lead to loss of genetic variation. Below I examine each of these arguments.

Inbreeding and its consequences on individual fitness will be covered in more detail later in this book. At this point it suffices to define inbreeding as matings between individuals that carry alleles identical by descent. In non-random mating populations, such as species that are fragmented into subpopulations with limited dispersal, the frequency of matings between individuals that carry alleles identical by descent (i.e. relatives) is increased. In diploid organisms this has the consequence that heterozygosity will be reduced. In a closed population of finite size, the rate at which inbreeding will increase as measured by the inbreeding coefficient is given by:

$$F_t = 1 - (1 - (1/2N))^t$$

where N is population size and t is the number of generations since the founding generation (Falconer and Mackay 1996). From this formula it can be seen that F will increase faster with small N and more slowly with large N (Fig. 1.1). It is important to note that inbreeding as such may not have any harmful effects. It is when inbreeding leads to inbreeding depression that endangered populations become severely impacted. I will come back to the issue of inbreeding and inbreeding depression in Chapter 3.

The random loss of alleles due to the stochastic processes of Mendelian segregation and sexual reproduction is more or less negligible in large populations. In large populations selection is the main cause for shaping allele frequencies. However, in small populations the importance of genetic drift becomes a far more important process. Assuming a biallelic locus subject to drift and selection, selection predominates when $4N_e s \gg 1$ (where N_e is the effective population size and $1-s$ is the fitness of homozygotes relative to the heterozygote) and drift predominates when $4N_e s \ll 1$ (Kimura 1983). From these inequalities it is evident that for any given level of selection it is more likely that drift becomes more prominent when N is small.

In general, the proportion of selectively neutral genetic variation lost per generation is $1/(2N_e)$. Small populations (low N_e) thus lose genetic variation faster than larger ones (Wright 1969). In real populations the actual population size N is always higher than N_e due to variance in the number of breeders and family sizes, fluctuations in population size, and unequal sex ratios (Wright 1969). Frankham

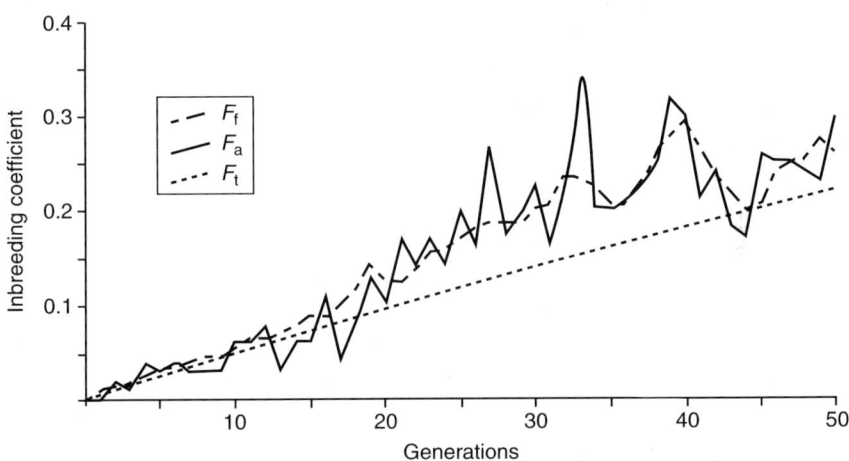

Figure 1.1 Inbreeding increases with time in a closed population. The line (F_t) is the theoretical expectation. The other trajectories (F_a and F_f) are based on stochastic simulation using Populus 5.3.

(1995) suggested that the ratio N_e/N in natural populations would typically be in the order of 0.1.

Large portions of the genome of any organism are selectively neutral, or at least nearly so at any given point in time. It may thus be argued that genetic variation is irrelevant for population survival. However, even if much of the standing genetic variation in an endangered population at any given point in time is selectively neutral, significant and important portions are not. Furthermore, standing genetic variation may be needed when and if conditions change. Alleles that are selectively neutral may become selectively advantageous in the future. Populations that have lost genetic variation have lost the ability to adapt to new conditions and consequently have become more prone to extinction.

To maintain levels of heritable variation in quantitative characters and ensure evolutionary viability, Franklin (1980) suggested a minimum effective populations size of $N_e = 500$. Taken together with the suggestion that a minimum population size of 50 is required to safeguard a population from extinction due to demographic stochastic reasons (Lande 1976), this has become known as the 50/500 'rule'. With $N_e/N = 0.1$ this would mean that the actual population size of any endangered population would need to be in the order of 5000 individuals. Clearly, many endangered populations typically harbour fewer individuals than this. Furthermore, it has been argued that since most genetic variation in quantitative characters in fact is harmful and maintained in the recessive state, only a fraction is quasi-neutral and potentially adaptive. This would increase the critical number to an N_e in the order of 5000 and the critical N to 50 000 (Lande 1995, 1999). If these theoretical considerations apply to real populations, genetic considerations are needed for many populations regardless of whether they are considered endangered or not.

Another harmful result of genetic drift is that drift may cause fixation of mildly deleterious mutations. Fixation of such mutations leads to a reduction in individual fitness which may negatively impact endangered populations. As shown above, drift is more potent in small populations and endangered populations tend to be small. Since accumulation of deleterious mutations speeds up as a population's size decreases, the population may be caught in a negative feedback loop towards extinction. This process has been termed mutational meltdown (Lynch et al. 1993). There is controversy over the significance of this process and its relevance to population persistence (see Gaggiotti 2003 for a review). The time scales involved when mildly deleterious mutations accumulate are in the order of hundreds of generations and their effect is only predicted to be severe in very small populations ($N < 100$; Lande 1999).

In empirical research it is often not possible to sort out the relative effects of inbreeding and drift since both processes work in the same direction, reducing genetic variation. A review of data from studies of plant species show that small and isolated populations typically harbour less genetic variation than large populations within dispersal distance of other populations of the same species (Fig. 1.2).

Both reduction of individual fitness and population adaptability ultimately lead to lower reproduction and increased mortality, factors that further lower an already small population size. When populations are caught in this downward spiral they are said to be trapped in an extinction vortex (Fagan and Holmes 2006) (Fig. 1.3).

1.3 Evidence from wild populations of a link between low genetic diversity and extinction

The extinction vortex hypothesis makes a few clear predictions as to whether genetic factors are important in the extinction of endangered species. The first prediction is that small and endangered populations and species should harbour less genetic variation as compared with taxonomically related non-threatened taxa. This prediction has been tested in an extensive meta-analysis of 170 threatened taxa and their non-threatened sister taxa (Spielman *et al.* 2004). The analysis covered both plants (Angiosperms and Gymnosperms) and animals (vertebrates and invertebrates). Average heterozygosity was lower in threatened taxa in 77% of the comparisons, a result which is significantly different from the null hypothesis of no difference between threatened and non-threatened taxa. On average, heterozygosity was 35% lower in threatened taxa than in non-threatened taxa. These results indicate lowered evolutionary potential, compromised reproductive fitness, and elevated extinction risk for threatened taxa. From this study it is clear that most taxa are not driven to extinction before genetic factors affect them negatively and furthermore that genetic methods in most cases can be employed to diagnose threatened taxa, at least when there is taxon we can identify *a priori* as non-threatened for comparison. The second prediction is that known cases of extinction should commonly be preceded by a radical loss of genetic diversity.

For obvious reasons it is not very common for species and populations that go extinct to have been extensively surveyed for genetic variation prior to their extinction. An exceptional case is the now-extinct heath hen *Tympanuchus cupido cupido* which once inhabited grasslands and barrens along the mid-Atlantic coast of eastern North America. This species was once numerous throughout its former range but went extinct on the mainland by around 1870. The last bird was seen on the island Martha's Vineyard on the 11 March 1932 (Johnson and Dunn 2006). Extraction of DNA from museum skins and subsequent amplification of mitochondrial DNA (mtDNA) has revealed that 30 years prior to their extinction, heath hens on Martha's Vineyard had low levels of mtDNA variation as compared with

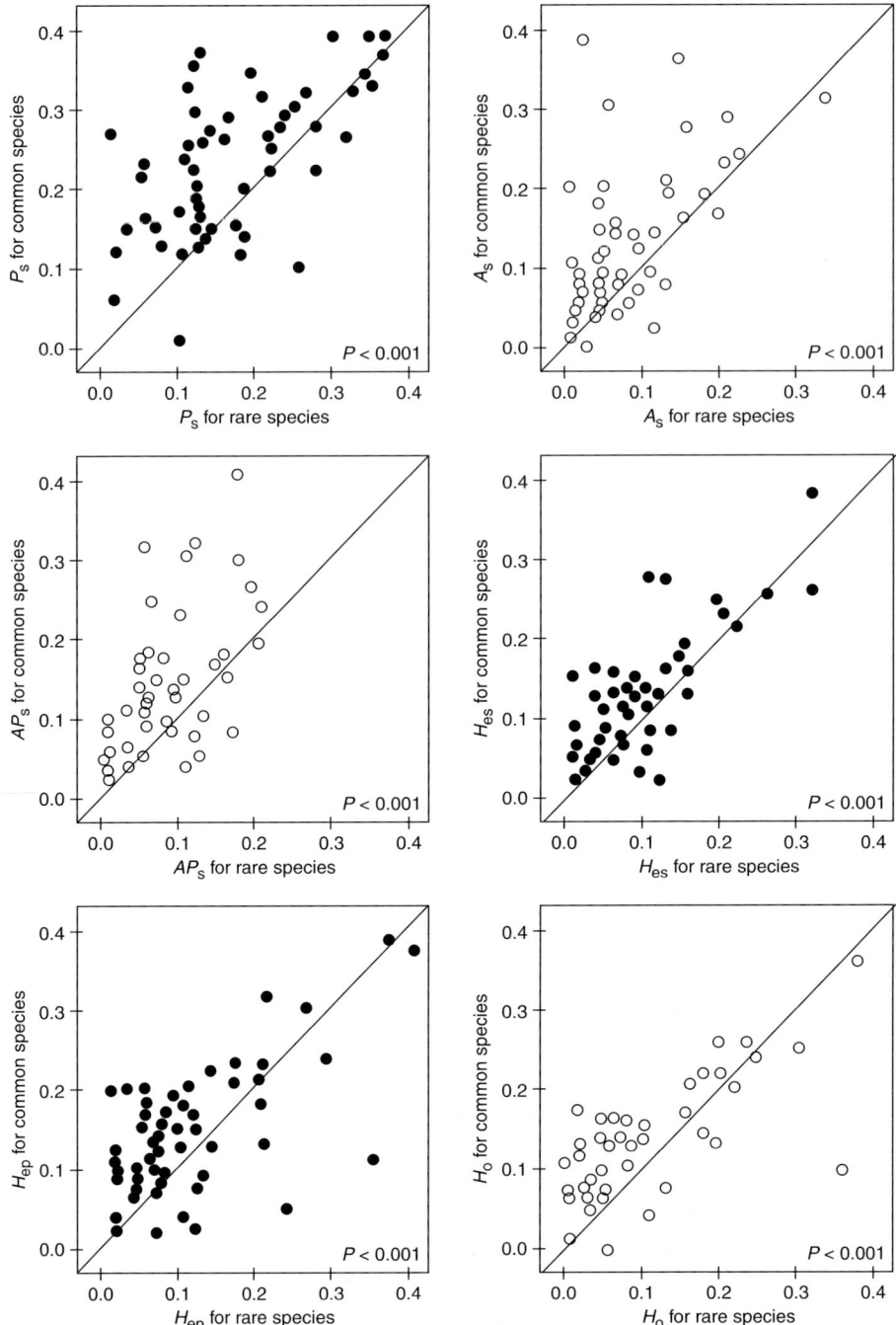

Figure 1.2 Levels of genetic (isozyme) variation in rare and common plant species. The line of equal expectation is drawn through each figure and *P* values are found in the right-hand corner of each graph. Subscript s indicates species-wide values, subscript p indicates the mean of population values. From top left to bottom right: *P*, percentage of polymorphic loci; *A*, alleles per locus; *AP*, alleles per polymorphic locus; H_e, expected heterozygosity; H_o, observed heterozygosity (from Cole 2003, reprinted with permission from the publisher).

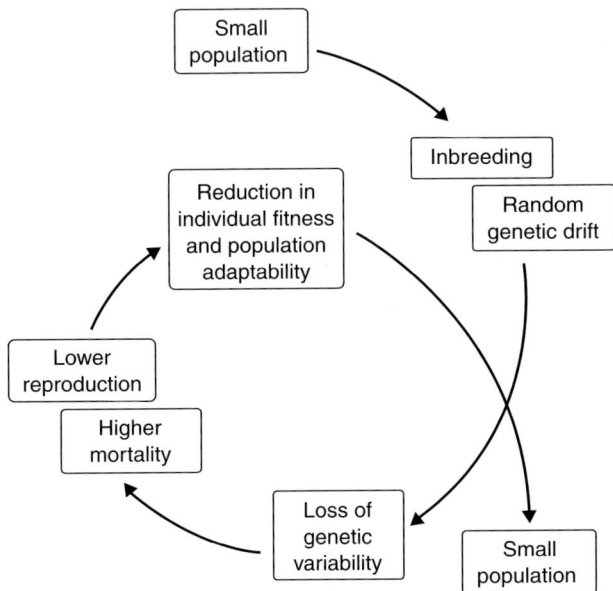

Figure 1.3 A schematic representation of the extinction vortex.

contemporary populations of prairie chickens (which are considered subspecies of heath hens, and all of which are considered presently endangered to varying degrees; Johnson and Dunn 2006).

The species extinction more or less coincided with the settlement of Europeans in North America. Approximately 200 years after the arrival of Europeans and colonization of the eastern United States, heath hens perished on the mainland. Thus it is more than likely that the extinction of heath hens were caused by human actions. Second, the heath hens on Martha's Vineyard indeed had exceptionally low genetic variation prior to their extinction (mitochondrial DNA haplotype diversity, $h = 0.363 + 0.029$; Johnson and Dunn 2006). Other endangered prairie chicken populations typically display a haplotype diversity in the region of 0.900. The only contemporary exception is the extremely endangered Attwater's prairie chicken *Tympanuchus cupido attwateri* which in museum samples from 1951 to 1954 had a haplotype diversity of 0.900, but presently (1998–2000) subpopulations lie in the range of 0.400–0.800, showing that the Attwater's prairie chicken is presently suffering loss of genetic diversity.

Habitat destruction, overexploitation by humans, disease, and poor reproductive success as a consequence of low genetic variation have all been cited as contributors to the decline and extinction of species including heath hens (Gross

1928, Simberloff 1998, Westemeier *et al.* 1998). Throughout this book I will argue that it is likely that all these factors contribute to the extinction of endangered populations: the argument for a role of genetics does not preclude other factors also being important. However, the reverse argument, that genetic factors may be considered less important, have indeed been put forward (Lande 1988, Caughley 1994, Elgar and Clode 2001). In the case of the heath hen I would personally bet on human overexploitation being the main reason for heath hen populations to become small and fragmented. This fragmentation ultimately led to a point when heath hen populations became vulnerable to loss of genetic variation. Whether or not the last heath hen population ultimately went extinct due to genetic effects we can never be certain. However, the last population did indeed show the diagnostics based on mtDNA data of being genetically impoverished. A prudent interpretation of these data is that a multitude of factors may contribute to the extinction of species. Very few, if any, numerous and widespread species go extinct without a period of range contraction, fragmentation, and severe contraction in numbers. A lot is gained in the preservation of biodiversity if populations can be diagnosed as threatened before genetic and demographic stochastic events lead to their extinction. Furthermore, if small and fragmented populations indeed commonly perish due to genetic reasons it is important to prevent this from happening by subjecting such populations to genetic restoration (Ingvarsson and Whitlock 2000, Ingvarsson 2002).

In the above example the ultimate reason for the extinction was unknown. Studies of populations that has nearly gone extinct but have been rescued may provide clues to the role of genetics in extinction. An example of such a species is the Scandinavian wolf. By the late twentieth century, the Scandinavian population of wolves *Canis lupus* had been almost driven to extinction. Only stray individuals persisted and there had been no successful reproduction reported for years. In Finland, however, a few reproducing packs remained. After many years without reproduction one pack in Sweden suddenly produced offspring in 1983, nearly 1000 km from the closest known packs in Finland and Russia (Liberg *et al.* 2005). The Swedish population has since been monitored closely but showed signs of inbreeding depression, such as hereditary blindness, known from captive populations (Laikre and Ryman 1991, Ellegren 1999). Detailed studies of a pedigreed population from 1983 to 2002 showed that the entire Scandinavian population was founded by only three individuals and that the inbreeding coefficient F varied between 0.00 and 0.41 for wolves born during the study period. First-winter survival of pups was strongly negatively correlated with their inbreeding coefficient ($r^2 = 0.39$, $P < 0.001$; Liberg *et al.* 2005). In 1991, the Scandinavian population started to increase and current numbers are now about 10–11 breeding packs annually, corresponding to about 100 wolves. It has been proposed that the sudden increase in numbers coincided with the immigration of a single successful breeder of Finnish or Russian origin in 1991 (Vilà et al. 2003). Vilà et al. suggested that of 72 wolves born after 1993, 68 can trace at least part of their ancestry back to this immigrant male. Thus, if correct, the genetic restoration of the Scandinavian wolf population is to a large extent due to one individual. In this case it seems clear that genetic effects cannot be ignored in conservation efforts (Ingvarsson 2002).

Another possible example of genetic rescue is an isolated population of adders *Vipera berus* at the very southern tip of the Scandinavian peninsula. This population suffered from low reproductive rates, possibly caused by inbreeding depression. Following the experimental movement of individuals to this population, reproductive rates has increased (Madsen *et al.* 1999). This suggests that enforced or natural low levels of migration between individuals of endangered populations can restore genetic diversity and reduce the risk of extinction, especially if the cause is inbreeding depression.

Yet another detailed study of possible genetic rescue is the greater prairie chicken *Tympanuchus cupido pinnatus* in midwestern North America. This once widespread species is now split into several disjunct ranges (Bouzat *et al.* 1998a). Especially in the eastern part of the range, in Wisconsin and Illinois, populations have been severely contracted and reduced in numbers. In Wisconsin the estimated population size was 54 850 birds in 1930 (Gross 1930). Since the 1950s the estimate has been around 1500 birds, a number observed also in 2003 (Bellinger *et al.* 2003). In Illinois greater prairie chickens declined from over 25 000 birds in 1933 to about 2000 in 1962 and 46 birds in 1994 (Westemeier *et al.* 1998). In Wisconsin, microsatellite allelic diversity has been shown to have been lost in the contemporary population compared to the historic population sampled from museum skins (Bellinger *et al.* 2003). In Illinois similar observations were made while no loss of alleles could be observed in the larger populations in Kansas, Minnesota, and Nebraska (Bouzat *et al.* 1998a, 1998b). Data from Illinois show that, with the exception of a temporary peak in male numbers in the early 1970s, displaying male numbers have steadily declined since the start of observations in 1963. Corresponding to this decline is a decline in the percentage of eggs hatched in observed clutches. Hatchability went down from a usually observed value of about 90–95% to around 65% by 1990 (Fig. 1.4). Following the translocation of birds in 1992, hatching success was restored to the usual level of around 95% (Westemeier *et al.* 1998). These data suggest that hatching success was impaired due to inbreeding depression and that genetic considerations cannot be ignored while attempting to rescue these endangered populations.

The previous examples have been on animals but the above-cited principles about genetic variation and extinction risk should also apply to plants and other organisms. Yet many botanists have been strong advocates for the case that genetic variation is of minor importance when studying extinction of endangered

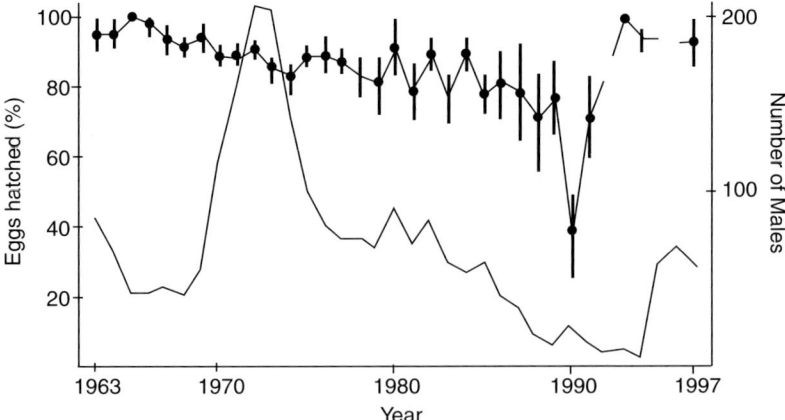

Figure 1.4 Annual mean hatching success (filled circles) of greater prairie chicken eggs and counts of lekking males (solid line) in Jasper County, Illinois, USA, in 1963–1997. Translocations of non-resident birds began in August 1992 (from Westemeier et al. 1998, reprinted with permission from the publisher).

populations. Holsinger and coworkers even went so far as to suggest that "changes in the genetic structure of plant populations are likely to threaten its persistence only if they involve loss of self-incompatibility alleles or genetic assimilation through hybridization with a reproductively compatible related plant species" (Holsinger et al. 1999). Thus genetic reasons for extinction were argued to be important only under rather extreme conditions. Yet a review of genetic variation in rare and common plant species showed that rare species have less genetic variation in almost all aspects measured, in accordance with the extinction vortex hypothesis. The review concluded that "rare plants evidently have more significant reductions in genetic variation and gene flow than have been recognised previously" (Cole 2003).

Oostermeijer and coworkers have been using both demographic and genetic approaches to plant conservation in the Netherlands (see references in Oostermeijer et al. 2003). The Netherlands may be a particularly relevant area of the world for learning about fragmentation and anthropogenic influence on wild species. The human population size of the Netherlands has increased and land use has changed dramatically over the last few centuries. Thus many native species have become fragmented and reduced in numbers: so-called "new rares" (see Huennecke 1991). Studies on Dutch new rares (Oostermeijer et al. 2003) show that there is indeed a relationship between genetic variation and population

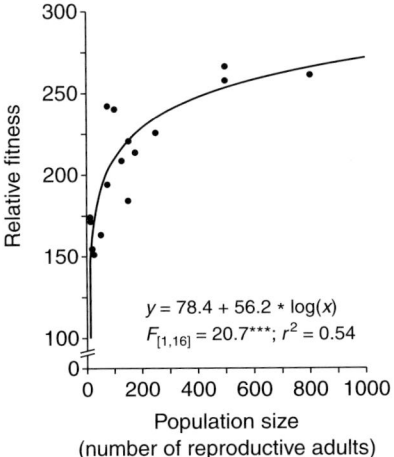

Figure 1.5 Relationship between relative fitness and population size in *Gentiana pneumonanthe* (from Oostermeijer *et al.* 2003, reprinted with permission from the publisher).

size such that smaller populations generally have less variation than larger ones. Studies also suggest that genetic variation is related to individual fitness in populations of Marsh gentian *Gentiana pneumonanthe* (Ooostermeiijer *et al.* 1995) and Leopard's bane *Arnica montana* (Luijten 2001) in the Netherlands and northern rockcress *Arabis petraea* (Schierup 1998) in Denmark. More heterozygous individuals perform better than less heterozygous ones, suggesting that inbreeding depression may be at work in these populations. If population size is related to genetic variation, the authors expected that there will also be a correlation between population size and fitness—related parameters. This has indeed been observed in *G. pneumonanthe* (Oostermeijer *et al.* 1994; Fig. 1.5), *A. montana* (Luijten *et al.* 2000), and spiked rampion *Phyteuma spicatum* (Boerrigter 1995). The studies on Dutch new rares also suggest that environmental stochasticity is important in understanding local extinction and the authors argue for an integrated approach where both genetic and demographic factors should be considered to preserve endangered plant populations.

All the above examples point to the direct genetic threat to endangered populations being mediated mainly via inbreeding depression and not so much due the stochastic loss of genetic variation or fixation of mildly deleterious alleles through genetic drift. I will soon return to a few examples of populations that seem to thrive despite the fact that they have been shown to be low in genetic variation but first there is a need to discuss a related issue. It has been proposed that inbreeding depression may not always be a consequence of inbreeding in endangered populations. One of the most famous examples is the case of the

Mauritius kestrel *Falco punctatus*. This population has been severely bottle-necked (contracted in numbers). The entire world population was down to one breeding pair in 1972; however, by 1994 there was more than 200 birds but no signs of inbreeding depression (Groombridge *et al.* 2000). This population is obviously inbred since all individuals are descendants of the same pair in the 1970s. One possible explanation is that during the severe bottleneck not only beneficial genetic variation was lost but also alleles that cause inbreeding depression. When the population became purged from these harmful alleles it could tolerate high levels of inbreeding without suffering from inbreeding depression.

It thus seems as though inbreeding may lead to inbreeding depression in some cases but not in others. A study of a fritillary butterfly species, *Melitaea cinxia*, by Saccheri and others (1998) hints at a possible solution as to why some species seem to tolerate inbreeding while others do not. In this study it was shown that local extinction risk is dependent on both ecological variables (mainly degree of isolation and population size) and genetic variation (Fig. 1.6). In particular, when ecological and genetic factors coincided, small and inbred populations became vulnerable to extinction. It was suggested that in the metapopulation system of this butterfly, the purging is not strong enough to deplete the system of the alleles responsible for inbreeding depression. The deleterious alleles would always remain in the heterozygous state in the large subpopulations that never go extinct. However, in small and inbred populations these alleles become expressed as homozygotes and cause inbreeding depression and ultimately population extinction. In a species like the Mauritius kestrel the deleterious alleles cannot 'hide' in a large population but will be exposed to selection and removed. However, populations like the Mauritius kestrel are more exposed to the risk that mildly deleterious alleles may become fixed through chance effects despite being selected against.

There are other examples of endangered species, which like the kestrel in the example above, seem to have low genetic variation and yet thrive and increase in population size. Norwegian red deer *Cervus elaphus* are comparable in microsatellite genetic variation with other threatened deer species that are signified by low genetic variation, yet the Norwegian population of red deer in recent years has expanded in number (J. Höglund and L. Kastdalen unpublished results). Another example is the Swedish beaver *Castor fiber* population which was founded in the 1920s by only a few individuals imported from Norway after being hunted to extinction in the late nineteenth century (Ellegren *et al.* 1993, Mikko and Andersson 1995). Today the Swedish population of beavers is expanding and numbers are now in the order of thousands of individuals. The list of similar examples can be made longer; for example, northern elephant seals *Mirounga angustirostrus* (Bonnell and Selander 1974, Hoelzel *et al.* 1993). A possible explanation is that purging may have provided a short-term opportunity for these endangered populations by allowing them to escape the threats of inbreeding depression.

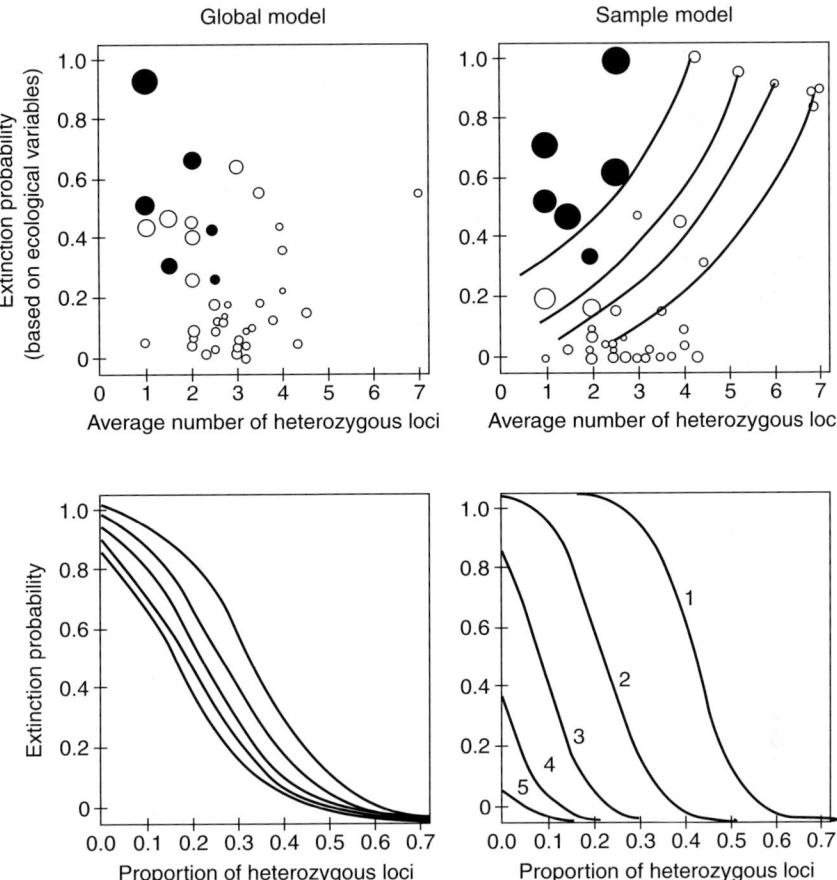

Figure 1.6 Inbreeding and extinction risk in the Glanville fritallary using two statistical models (from Saccheri *et al.* 1998). Upper panels: the probability of extinction predicted by the models without heterozygosity (extinct populations are shown by black circles and surviving populations with white circles). The probability of extinction predicted by the full model, including heterozygosity, is proportional to circle size. For the sample model, appropriate isoclines for the extinction risk predicted by the model, including ecological factors and heterozygosity, are drawn. The lower panels show the relationship between the risk of local extinction and heterozygosity predicted by the two models. Model predictions are shown for local population sizes of one to five larval groups (reprinted with permission from the publisher).

However, genetically impoverished populations may face inescapable threats in the long term. The Scottish population of capercaillie *Tetrao urogallus* became extinct around 1790. Restocking started in 1835, when 65 birds were imported from Sweden. From 1930 to 1970 numbers were estimated to have fluctuated just above 20 000 individuals, suggesting that the species had reached its local carrying capacity. However, since the mid-1970s numbers have plummeted to around 2000 despite the fact that, if anything, the forest habitat in which the species lives has increased. There is yet no firm evidence that Scottish capercaillie have been reduced in numbers for genetic reasons. However, genetic variation in Scottish capercaillie is indeed lower than in other parts of the species' range (but not as low as in the Pyrenees and the Cantabrian mountains) and thus it is possible that the low genetic variation due to a founder event may have contributed to the decline in population size (S. Piertney, personal communication).

Another species with a similar history is the American crayfish *Pacifastacus leniusculus*, native to northwestern USA and southwestern Canada, imported to Sweden during the twentieth century because it is resistant to disease caused by a fungus, *Aphanomyces astaci*, which is lethal to European crayfish, *Astacus astacus*. The fungal disease is of North American origin and the American crayfish and the fungus have a long evolutionary history and therefore the American crayfish is tolerant to the disease. Ironically, in Swedish waters the main agent spreading the disease is the imported American crayfish, causing massive extinction of the native species. Furthermore, there is evidence that American crayfish are superior competitors and often exclude European crayfish when living in the same waters. Altogether, introducing American crayfish to Scandinavia has not been a good idea. It is possible, but to my knowledge unknown, that American crayfish lost genetic variation during the founder event when they were introduced to Sweden. What seemed to happen while this book was being written was a crash of populations of American crayfish in Sweden (Söderhäll 2004), which may give native European crayfish a second chance. American crayfish may be yet another example of a species that after an introduction and low numbers with accompanying low genetic variation fared well for a while. However, in the long run genetic variation was too low to safeguard against new threats. More research is needed on both Scottish capercaillie and Swedish American crayfish to test whether this hypothesis is true.

1.4 **Experimental studies**

There have been a few experimental studies to test whether inbreeding and/or reduced levels of genetic variation leads to greater extinction risk. Indeed, the rate of extinction for small and/or inbred experimental populations appears to be greater than for large populations (Latter *et al*. 1995, Frankham 1996, Newman and Pilson 1997, Bryant *et al*. 1999, Reed and Bryant 2000, Reed *et al*. 2003).

Using the housefly *Musca domestica*, Reed and Bryant (2000) compared fitness and rates of extinction among populations kept either at constant effective population sizes of 50, 500, or 1500 or passed through bottlenecks reducing N_e

to five individuals. The results demonstrated that population fitness, measured as larval viability, total eggs, and total progeny, was closely related to population size. Within six generations small populations maintained at an effective population size of 50 individuals were significantly lower in all three fitness measures than larger populations. The loss of fitness decreased the longevity of the small lines with five out of six lines going extinct by generation 64. Similar results were obtained in another experiment (Bryant *et al.* 1999). Taking the two experiments together, predicted extinction times (based on the regression of viability on number of generations) were under 100 generations for an effective population size up to 100 and increased to over 400 generations when N_e was 500 and above.

Another aspect of this experiment was that in the founder-flush treatment, when populations were bottlenecked to $N_e=5$ and then allowed to grow to approximately 2500 individuals in seven generations, lines exhibited some recovery in larval viability after the initial bottleneck (see also Bryant *et al.* 1990). This suggests that these lines may have been purged for alleles causing inbreeding depression, this echoing the explanation for why the falcons on Mauritius, cited above, may survive severe inbreeding. However, the purged lines did worse under dietary and thermal stress. The authors suggest that whereas a bottlenecked population may adapt to a particular environment its adaptability may be low and suggest that the lack of adaptability may outweigh any benefits of bottlenecks due to purging (Reed and Bryant 2000).

Studies of the evening primrose, *Clarkia pulchella*, further suggest that inbred populations run higher risks of extinction. In experimental populations that all had the same number of founders but which differed in the relatedness among founders, inbred populations were more prone to extinction (Newman and Pilson 1997).

Experimental studies using the fruit fly *Drosophila melanogaster* have attempted to examine the relative roles of inbreeding and population size on cumulative extinction rate (Reed *et al.* 2003). Survival dropped faster with increasing levels of inbreeding at low effective size treatment ($N_e=2.6$) than in any of the treatments with larger effective size ($N_e=10$ and 20, respectively; Fig. 1.7). For any given level of inbreeding extinction was greater for the lowest N_e. This result may imply that the slower the inbreeding (larger N_e) the more effective the purging of deleterious alleles. However, the authors are cautious of such an interpretation, mainly owing to the fact that both the treatments with a higher N_e (those that are predicted to be purged) had lower survival of lines than outbred controls. Thus purging was not considered to have removed all deleterious alleles causing inbreeding depression. It has been concluded that purging is generally inefficient in reducing inbreeding depression (Allendorf and Ryman 2002). Other experiments have shown that inbred populations have a significantly higher short-term probability of extinction than non-inbred populations (Bijlsma *et al.* 1999, 2000).

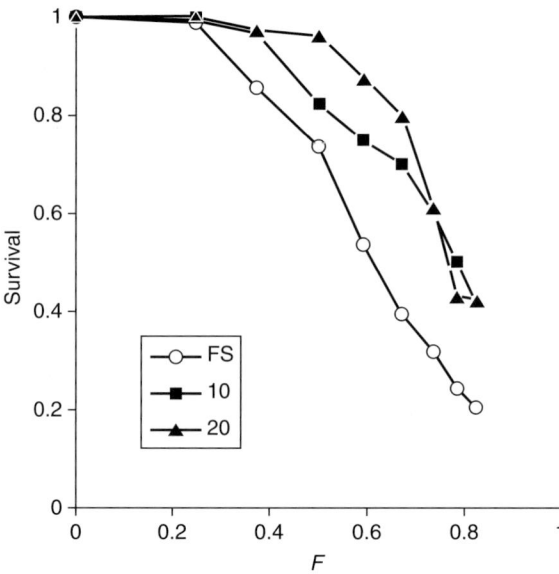

Figure 1.7 Cumulative extinction rate plotted against inbreeding coefficient, F, for three experimental population size treatments of D. melanogaster: $N_e=2.6$ (FS), $N_e=10$ (10), and $N_e=20$ (20) (from Reed et al. 2003, reprinted with permission from the publisher).

Moreover, the negative effects of inbreeding became enhanced under stressful environmental conditions. These results indicate that inbreeding and environmental stress interact synergistically and make small populations vulnerable to extinction.

Survival was negatively affected by environmental stress such that survival decreased for any given level of inbreeding when populations were subjected to differential treatments of environmental stress (Reed et al. 2002). This again suggests that the detrimental effects of inbreeding are environmentally dependent (Armbruster and Reed 2005). Since threatened populations often live in stressed and marginal habitats it is therefore predicted that the negative effects of inbreeding may be exaggerated in such cases. In experiments using the amphipod Gammarus duebeni, survival did not differ when comparing stressed treatments and benign laboratory treatments using outbred lines (inbreeding coefficient $F=0$). However, inbred lines ($F=0.25$) experienced reduced survival under stressful field conditions (Gamfeldt and Källström 2007). That inbreeding depression is environmentally dependent shows that, in conservation biology, genetic studies cannot be isolated from ecological studies. The genetics need to be put in an ecological and demographic perspective to increase our understanding of the factors that may cause population extinction and biodiversity loss.

1.5 Conclusions

It appears that many studies of genetic causes for extinction seem to suggest that inbreeding depression is the main genetic problem in conservation biology. On the other hand, hardly any study has convincingly shown that reduced adaptability or fixation of mildly deleterious alleles have contributed to extinction. It may therefore seem prudent for conservation geneticists to focus on inbreeding and inbreeding depression. However, as has been hinted at in studies of wild endangered species and shown in a few experimental studies, such a conclusion may be premature. Documenting cases in the wild when inbreeding can be excluded as a factor is extremely unlikely, owing to the fact that both loss of alleles and inbreeding lead to population extinction and that their relative effects may be coupled (effects of inbreeding becoming exaggerated at low N_e). Furthermore, effects of lost adaptability may only be discernable in the long run; that is, on time scales beyond the scope of research projects or even the life times of researchers. Both loss of alleles and inbreeding can be treated with the same cure: transplantations from conspecific populations that aim to restore and maintain genetic variability in the threatened populations. However, as will be discussed in later chapters, such transplantations are not always uncontroversial.

2 *How to measure genetic variation*

From the examples in the previous chapter it is obvious that there are many ways to assay and analyse genetic variation. Choice of analytical method is partly dependent on the type of genetic marker used. Furthermore, different aspects of variation that can be assessed depend on whether the marker is subjected to selection (non-neutral) or not (being selectively neutral). Here I outline the most common measures of genetic variation used in conservation genetic studies (see e.g. Karp *et al.* 1997). I have chosen to structure this discussion around the type of data collected and a summary of the different markers used can be found in Table 2.1.

2.1 Codominant neutral variation

Genetic variation in endangered species is most commonly assayed using genetic markers that are suspected to be neutral or nearly neutral, such as allozymes, microsatellites, and—increasingly—neutral single nucleotide polymorphisms (SNPs; see below). These are all codominant markers, meaning that in diploid genomes there are two copies at any locus. By neutral we mean that there is no evidence of selection being involved in shaping the allele frequencies observed at the loci studied. This is most often assessed by testing whether the allele frequencies differ from what is expected from Hardy–Weinberg expectations. The Hardy–Weinberg equilibrium expectation is the heterozygosity expected at a locus given that the alleles observed in a sample segregate randomly according to Mendelian inheritance.

Genetic variation at allozyme (or isozyme) loci are assayed at the protein level using starch gel electrophoresis and used to be the marker of choice in early studies of genetic variation. Allozyme variation studies are still performed but have become less common. Contemporary studies instead tend to use microsatellite variation as an alternative when studying genetic variation of endangered species. SNPs are still not common when non-model organisms are studied.

There are basically two reasons for the change from allozymes to microsatellites. First, allozyme variation has sometimes been suspected to be non-neutral, meaning that at least some of the variation observed within and among populations may be attributed to selection (e.g. Szarowska *et al.* 1998). However, the same argument has been suggested to apply also to microsatellites (Kauer *et al.* 2003) and hence it is a poor reason for choosing microsatellites instead of allozymes as the marker of choice in any study. However, the second reason, destructive sampling, is more relevant. Allozyme studies require larger amounts of high-quality tissue and most often involve culling the study organisms. Clearly, culling is not a good idea when studying endangered species. Even if enough material can be collected without culling, preservation of the tissue until relevant material can be extracted in the laboratory is much more cumbersome in the case of allozymes compared with microsatellites. One aspect in favour of allozymes is the relatively low laboratory costs involved given that sufficient material can be obtained.

SNPs have become increasingly popular in genetic studies of model organisms. The most often cited reason is that, in contrast to microsatellites, the mutational processes involved in creating a SNP is simple and well understood. Microsatellites are believed to evolve primarily because of slippage of the endogenous DNA polymerase during transcription but other mutational processes may also be involved that complicate analyses and interpretations (Eisen 1999). According to the stepwise mutation model, new microsatellite alleles are created by addition or removal of repeat motifs. This is thought to occur relatively commonly (mutation rates in the order of 10^{-3}). This has the consequence that any allelic state may have arisen in the evolutionary past of a study population more than once. In contrast, SNPs are believed to evolve primarily due to point mutations and/or via insertions and deletions, events that occur much more rarely (in the order of 10^{-6} per generation). Thus any two SNP alleles can safely be assumed to be traced back to a unique mutational event which greatly simplifies the theory for understanding the patterns of genetic variation in contemporary populations and the tools used to analyse such patterns.

For allozymes, microsatellites, and SNPs many of the analytical tools for studying genetic variation are the same. The following metrics occur in the literature.

2.1.1 Percentage of polymorphic loci

This may appear a straightforward measure but different studies vary in what criteria are used for scoring a locus as polymorphic. A locus could be defined as monomorphic if the most common allele frequency is 100, 99, or 95% of all sampled alleles. Since loss of rare alleles is expected to be one of the most immediate results of reduced population size, either the 100 or 99% criterion may be better estimates in endangered species.

Table 2.1 List of genetic markers and comments on their use and feasibility in studies of genetic variation (from Krutowskii and Neal 2001). Most often these markers are assumed to assay neutral DNA variation.

Feature	RFLP	Microsatellites	RAPD	AFLP	Isozymes
Origin	Anonymous/genic	Anonymous	Anonymous	Anonymous	Genic
Maximum theoretical number of possible loci in analysis	Limited by the restriction site (nucleotide) polymorphism (tens of thousands)	Limited by the size of genome and number of simple repeats in a genome (tens of thousands)	Limited by the size of genome, and by nucleotide polymorphism (tens of thousands)	Limited by the restriction site (nucleotide) polymorphism (tens of thousands)	Limited by the number of enzyme genes and histochemical enzyme assays available (30–50)
Dominance	Codominant	Codominant	Dominant	Dominant	Codominant
Null alleles	Rare to extremely rare	Occasional to common	Not applicable (presence/absence type of detection)	Not applicable (presence/absence type of detection)	Rare
Transferability	Across genera	Within genus or species	Within species	Within species	Across families and genera
Reproducibility	High to very high	Medium to high	Low to medium	Medium to high	Very high
Amount of sample required per sample	2–10 mg DNA	10–20 ng DNA	2–10 ng DNA	0.2–1 µg DNA	Several milligrams of tissue
Ease of development	Difficult	Difficult	Easy	Moderate	Moderate
Ease of assay	Difficult	Easy to moderate	Easy to moderate	Moderate to difficult	Easy to moderate

Automation/multiplexing	Difficult	Possible	Possible	Possible	Difficult
Genome and QTL mapping potential	Good	Good	Very good	Very good	Limited
Comparative mapping potential	Good	Limited	Very limited	Very limited	Excellent
Candidate gene-mapping potential	Limited	Limited	Useless	Useless	Limited
Potential for studying adaptive genetic variation	Limited	Limited	Limited	Limited	Good
Development	Moderate	Expensive	Inexpensive	Moderate	Inexpensive
Assay	Moderate	Moderate	Inexpensive	Moderate to expensive	Inexpensive
Equipment	Moderate	Moderate to expensive	Moderate	Moderate to expensive	Inexpensive

AFLP, amplified fragment length polymorphism; QTL, quantitative trait locus; RAPD, randomly amplified polymorphic DNA; RFLP, restriction fragment length polymorphism.

2.1.2 Alleles per locus/allelic richness

This measure obviously depends on sample size, so to compare samples of different sizes the number of alleles per locus is often replaced by allelic richness. Allelic richness is the number of alleles per locus rarefied to match the number of observations in the population with the lowest sample size (El-Mousadik and Petit 1996).

2.1.3 Expected heterozygosity

This is often also referred to as gene diversity and is the heterozygosity expected in a population given the observed allele frequencies. It is defined as $H_e = 1 - \Sigma p_i^2$, where p_i is the frequency of the ith allele at a locus. Almost invariably the mean over the number of loci is reported.

2.1.4 Observed heterozygosity

Observed heterozygosity is the mean of the observed proportions of heterozygotes, H_o.

2.1.5 Inbreeding coefficient

Observed and expected heterozygosity at a locus in a population may differ for a number of reasons. One is that inbreeding leads to more homozygous offspring than expected under Hardy–Weinberg (Conner and Hartl 2004). Assuming that inbreeding is the sole reason for deviations from Hardy–Weinberg expectations, the average inbreeding coefficient in a population can be estimated as $F_{IS} = (H_e - H_o)/H_e$.

2.1.6 Population differentiation

Wright (1929, 1951, 1969) was the first to note that in a species subdivided into more than one subpopulation, matings are non-random when considering the whole species. Thus even if matings are random within populations, subdivision causes a form of inbreeding when considering the whole species. The extent of population differentiation may thus be regarded as an inbreeding coefficient entirely due to population subdivision and in its most general form it is defined as $F_{ST} = (H_T - H_S)/H_T$, where H_T is the heterozygosity in all populations and H_S is the mean heterozygosity in the subpopulations. There is a rich and extensive literature on how to interpret and calculate F_{ST} (see Chapter 3 in this volume). In conservation studies F_{ST} is particularly relevant since in small populations drift is expected to increase when population size becomes small. Therefore one may expect populations of endangered species to show more subdivision than more numerous species. Often the differentiation between pairs of populations within a larger sample is calculated as pair-wise F_{ST}. In this case H_T is calculated for the combined sample of the two populations compared.

2.1.7 Gene flow

Assuming an island model—that is, with all populations equidistant from one another—and that the populations are of roughly equal size, Wright showed that the number of migrants per generation (the product of population size, N, and the probability of migration, m) is inversely related to population differentiation such as $F_{ST} = 1/(1 + 4Nm)$. Thus, if one is willing to accept the assumptions, with knowledge of allele frequencies in the populations F_{ST} can be calculated and consequently Nm may be derived. Because the assumptions necessary to derive Nm from allele frequencies are hardly ever met a measure of gene flow based on the mean frequency of private alleles (alleles unique to a single subpopulation) has been developed (Slatkin 1985, Slatkin and Barton 1989).

2.2 Dominant neutral markers

With codominant markers, the investigator can infer the state of each of the alleles at a locus and to directly infer the level of heterozygosity. Several methods (see below) have been developed where the researcher cannot directly infer heterozygosity of the 'alleles' detected by the marker, often referred to as dominant markers in line with the fact that if there is complete dominance at a locus, the allelic state cannot be inferred from the phenotype.

One of the first of these methods that used the PCR technique was randomly amplified polymorphic DNA (RAPD). With this method the researcher uses short (10–12 bp in length) primers that anneal randomly to the target DNA and amplify the DNA positioned between any two random primer pairs. If the primers anneal to the template DNA and if the targeted DNA sequence is short enough, an amplification product will be produced that can be visualized: a so-called RAPD profile of the targeted organism.

The advantages of RAPDs are that the technique does not require any knowledge of the targeted DNA and that it is relatively cheap. Among the disadvantages are that the technique is very sensitive to laboratory conditions and the quality of the DNA template used. Therefore the presence or absence of an amplification product could be because of differences among the targeted DNA sequences (the desired condition) or simply because samples differ in DNA quality or quantity.

Using restriction fragment length polymorphisms (RFLPs) the investigator may also detect dominant genetic variation within and between populations. This method takes advantage of the fact that restriction enzymes (restriction endonucleases) may cut DNA at specific target sequences throughout the genome depending on the enzymic system used. Different RFLP profiles are produced depending on whether a specific target sequence is present and the incidence of insertion/deletions and crossing-over events. RFLP profiles are usually enriched and visualized using Southern blots but other techniques are also available.

The advantage of using RFLPs is that it is a cheap, straightforward technique that like RAPDs requires no previous knowledge of the target DNA sequence (restriction sites are present in all organisms). It is considered a more reliable and reproducible technique than using RAPDs. On the downside, the investigator needs high concentrations of high-quality DNA and the laboratory protocol is often labour-intensive. Furthermore, the RFLP bands on a gel are not always easy to interpret, even with family data. For this reason, RFLP studies of population data are seldom conducted and it is not often used to assess genetic variation in endangered populations.

AFLP stands for amplified fragment length polymorphism and is a method that is akin to RFLP. Like with RFLPs, restriction enzymes are used to cut genomic DNA. This step is then followed by ligation of complementary double-stranded adaptors to the ends of the restriction fragments. The restriction fragments are amplified using primers complementary to the adaptor and restriction-site fragments and visualized (see Bensch *et al.* 2002, Vos *et al.* 1995).

AFLP is considered a more reproducible technique than using RAPDs and has become a popular technique to assess genetic variation, especially in non-model organisms since it also does not require any knowledge of the targeted DNA sequences. Since AFLPs use a PCR step, the required amount and quality of DNA is less than in RFLP studies.

As indicated, the techniques briefly outlined above produce data about dominant markers and therefore many of the metrics reviewed at the start of the chapter, such as heterozygosity, cannot be calculated directly from such data. However, various assumptions can be made in which dominant data can be interpreted and compared with the traditional measures. For example, under the assumption that the presence or absence of a restriction fragment corresponds to a genetic locus, allele frequencies can be estimated as q, equal to the square root of the frequency of '0' phenotypes (Lynch and Milligan 1994). There are also methods for estimating nucleotide and haplotype diversities (see below; see also Nei and Tajima 1981, Nei 1987).

2.3 Sequence variation

With sequence data selected (non-neutral) variation is most often detected and studied in the exons of protein-coding genes if the substitution has altered the amino acid sequence and biochemical properties of the encoded proteins. Within exons of protein-coding genes, such substitutions are called non-synonymous. However, non-neutral variation is also present in other parts of the genome, such as in control regions, and enhancer or promoter regions that bind to transcription factors, if such substitutions have phenotypic effects that may be affected by natural selection.

Silent or synonymous mutations are genetic changes that do not have any phenotypic effects. For example, if a mutation within an exonic region of protein-coding gene does not change the amino acid sequence of a protein such a mutation is referred to as a synonymous substitution. Silent mutations may also occur in non-coding DNA, such as in introns and pseudo-genes.

Within protein-coding exons, synonymous substitutions may occur because of the redundancy of the genetic code. The genetic code is read in triplets of nucleotides (called codons). Some codon positions are degenerate; that is, some nucleotide substitutions do not alter the amino acid sequence. For example, the third codon position may be fourfold degenerate, so the same amino acid will be encoded no matter what nucleotide is found in that position. Silent mutations are by definition evolutionarily neutral.

The most common measures of genetic variation with sequence data are described below.

2.3.1 Proportion of variable sites

This is calculated by counting the number of variable, segregating, sites, S, among the sampled sequences and dividing by the total number of sites, N, such as

$$p_n = S/N$$

The variance of this estimate can be obtained by

$$V(p_n) = (p(1-p))/N$$

(Nei and Kumar 2000).

2.3.2 Nucleotide diversity

This is the average number of nucleotide differences per site between any two randomly chosen sequences from a sample population (Nei and Li 1979, Nei 1987):

$$\Pi = \Sigma x_i x_j \pi_{ij}$$

where x_i and x_j are the frequencies of the ith and jth sequences and π_{ij} is the proportion of different nucleotides between sequences i and j. In randomly mating populations this corresponds to heterozygosity at the nucleotide level, which can be estimated by

$$\pi = N/(N-1)\Sigma x_i x_j \pi_{ij}$$

where N is the number of sequences. Formulae for obtaining the variance can be found in Nei (1987) and a resampling approach is described in Nei and Kumar (2000).

2.3.3 Haplotype diversity

A haplotype is a contraction of the phrase haploid genotype and is a stretch of DNA that is inherited as a unit. Thus the haploid mitochondrial DNA is an example of a haplotype since it is usually inherited as a single linkage group. In diploid genomes haplotypes are a set of closely linked nucleotides present on a chromosome that are inherited together. Thus haplotypes are stretches of DNA in linkage disequilibrium that are not broken up by recombination.

Haplotype diversity is defined as $1 - \Sigma f_i^2$ where f_i is the frequency of the ith haplotype. The reader will notice that this is the same formula as for expected heterozygosity for a codominant marker.

2.4 Non-neutral markers and neutrality tests

The same metrics as above may of course also be applied to genetic markers that have been subjected to selection. However, the researcher needs to be aware of the fact that the interpretation of the metric may be different in this case. For example, it is not possible to infer levels of inbreeding and migration if selection has been involved in shaping the allele frequencies at a given locus. Nevertheless, comparisons among neutral and non-neutral loci may allow other interesting inferences. As an example, in a recent study of an endangered bird species, the great snipe (*Gallinago media*), F_{ST} was compared within and among two geographic regions for microsatellites (neutral) and major histocompatability *Mhc* genes (non-neutral; Ekblom *et al.* 2007). It was shown that regional differentiation was more pronounced for *Mhc* genes than microsatellites. This may suggest that the snipe are differentially adapted to a local parasite fauna.

A number of tests for investigating whether any particular locus is evolving under neutral expectations or is under selection have been proposed in the literature. First and foremost standard tests for deviations from Hardy–Weinberg equilibrium may give a hint as to whether the locus is neutral or not. There may of course be reasons other than selection if a locus departs from neutral expectations, but this is a first test.

A commonly employed test with sequence data is to calculate the ratio of non-synonymous (dN) to synonymous substitutions (dS): dN/dS (or kN/kS). Purifying stabilizing selection will cause a low dN/dS ratio whereas diversifying positive selection will cause a high ratio. This aids in identifying genes or stretches of DNA that are evolutionarily constrained (low dN/dS) or, alternatively, codons that have been selected to be more variable than expected under neutrality (high dN/dS). The latter may apply to genes for immunity that sometimes may be under frequency-dependent selection (see Chapter 5).

Tajima's D is a statistical test designed to distinguish between DNA sequences evolving under neutrality and those evolving under a non-random process, including directional and diversifying selection. Note that expanding and contracting past population sizes and selection on genes located nearby in the genome (hitchhiking) may also cause deviations from neutral expectations.

The test rests on the assumption that under neutrality and in a population at mutation–drift equilibrium the expectation of nucleotide diversity $E(\pi) = \theta_\pi = 4N\mu$ (Kimura 1969). Another measure of nucleotide diversity is given under the an infinite-sites model, where the expectation of the number of segregating sites S is $E(S) = a\theta$ where $a = (\Sigma^{n-1} 1/i)$. Thus $\theta_S = S/a$. If the sequences evolve under neutrality the different estimates of θ should yield the same value. However, selection, or any other non-random process will differently change the values of S and π. The tests calculate $d = \theta_\pi - \theta_S$ and then

$$D = d/(V(d))^{1/2}$$

Under purifying selection D tends to be less than 0. However, under balancing selection, such as under heterozygote advantage, $D > 0$. As noted above D tends to deviate from the neutral expectation of 0 also under various demographic scenarios.

In other tests for evidence of selection, like the related Hudson–Kreitman–Aguadé (HKA) and McDonald–Kreitman tests, there needs to be sequence data from two different genes in at least two different species (Hedrick 2000). Therefore these tests are not so useful for studies of species that are facing extinction where genetic data may be scarce and there may not always be an obvious sister species for comparison.

2.5 Quantitative additive genetic variation

In the age of whole-genome sequencing of more and more organisms it is easy to forget more traditional methods to study quantitative genetic variation. There is currently somewhat of a renaissance in quantitative genetics due to new theory and software development. Furthermore, the combination of quantitative genetics and genomic tools are creating new research possibilities (Chapter 7). The ultimate goal of many genomic studies is to understand the genetic basis behind complex traits such as morphology, life-history variation, and disease resistance/susceptibility. By mapping the genomic regions associated with quantitative variation and linking genomic variation with phenotypic variation a fuller understanding of phenomena such as local adaptation to a variable environment, disease resistance, and life-history variation may be gained (see Chapter 7). However, the study of the genetic basis behind quantitative variation has a history which goes far beyond the discovery of the DNA molecule as the basis for inheritance. In the following I will briefly outline the theory of quantitative genetic variation of relevance to conservation genetics.

In the simplest case, all alleles at any given locus contribute equally to the phenotype determined by that locus. This is referred to as additive gene action. The alternative is dominance: that one allele contributes more (or less) than an equal share to the phenotypic variation. The simplest additive genetic model assumes that (1) all differences between individuals in a population are genetic, (2) alleles act additively (the alternative being dominance), and (3) epistasis can be ignored; that is, there is no interaction among genes. Thus the phenotypic value, P, of any given trait can be found by adding the genotypic values, G, for each of the alleles present at a locus. For example, if allele 1 has a G of 1 and allele 2 has a value of 2, then P is 3.

In reality, no trait is solely determined by its genotypic values. There is almost invariably environmental influence, E. A more realistic model is thus that the phenotypic value is the sum of the genotypic values and the environmental influence ($P = G + E$). Furthermore, it is unrealistic to assume that there is no dominance, D, and further that there is no interaction among loci, I. Thus the genotypic values, G, are influenced both by dominance and epistasis, such as $G = A + D + I$, where A stands for the additive effects. This means that a full quantitative genetic model is composed of $P = A + D + I + E$ (Fig. 2.1).

It is often stated that quantitative genetics is concerned with the analysis of complex traits affected by many genes. Whereas polygenic inheritance is by far

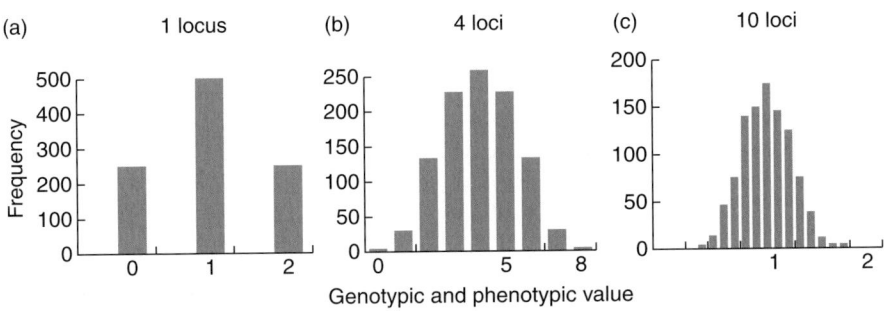

Figure 2.1 Outcomes of polygenic inheritance assuming two alleles per locus, A contributing 1 to the expression of the trait (genotypic and phenotypic value), and a contributing 0. Allele frequencies are 0.5 per locus. Trait expression is only due to genes.

the most common case for any real trait, the above example should have made it obvious that the simplest quantitative genetic case only involves a single locus and that the inheritance pattern is no different from a so-called Mendelian trait. If the environmental influence on P is low and if only one locus is involved, the phenotypic values will be distributed like in Fig. 2.1a. Figures 2.1b and 2.1c illustrate the effect of many loci being involved. The more loci, the more bell-shaped the distribution of phenotypic values.

Just as the phenotypic values of a trait can be partitioned into additive, dominance, and environmental components, the variance of the same trait in a population can be partitioned accordingly such as:

$$V_P = V_G + V_E = V_A + V_D + V_I + V_E$$

It is important to note that the important component in conservation studies (and indeed in any evolutionary application) is V_A since this is the only component that is inherited from the parents that can respond to selection. While it is true that individuals in a sense inherit the dominance at a given locus from their parents because they inherit their alleles at any given locus, any individual cannot inherit the dominance deviation at that locus nor any particular epistatic variation. Thus the V_D and V_I components are effects of the Mendelian lottery and V_A is the critical evolutionary component.

Heritability is defined as $h^2 = V_A/V_P$ and is the proportion of the variance in a trait that is due to additive genetic effects. Or put in another way, it is the proportion of the genetic variance that is heritable and which can respond to selection (see below). This is a dynamic property that is population-specific and subject to change throughout the evolution of a species. For example, under circumstances when the environmentally induced variance is high, heritability is lower.

Estimating heritability in natural populations is usually done either via parent–offspring regressions or sib analyses. In parent–offspring regressions, the offspring's value of any given trait is regressed on the parents' values (Fig. 2.2). Heritability is estimated as the slope of the regression between parents and offspring. The slope is multiplied with the inverse of the probabilities of identity by descent to obtain the heritability depending on what kind of comparison is made. For example, in the case of offspring on one parent, $h^2 = 2$ multiplied by the slope of the regression.

Sometimes it is impractical or impossible to estimate heritability via regression techniques. Under such circumstances it is better to use one or several experimental half-sib designs to estimate h^2. These designs have the further advantage of allowing several estimates of heritability and allow estimation of confounding effects such maternal effects. Furthermore, such analyses are the

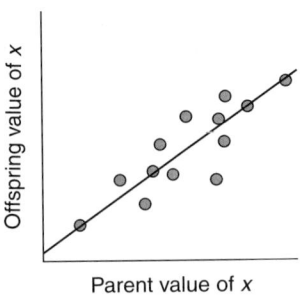

Figure 2.2 Hypothetical parent–offspring regression illustrating the heritability of a trait.

only choice when it is difficult or impractical to measure the same trait in parents and offspring. This is the case when offspring are not fully grown at the time of the measurement.

There are several techniques and protocols on how to perform sib analyses. For example, in a North-Carolina Design 1 experiment, males (sires) are mated with several females (dams). The offspring are raised and measured and the measurements subjected to analysis of variance (ANOVA) to estimate the variance components (Table 2.2). This allows three different heritability estimates: h^2_s, h^2_d, and h^2_{s+d}. The latter two contain both dominance and maternal effects while the former contains no dominance effects (see the table for definitions and details). By comparison the dominance component can be estimated.

Morphological traits of vertebrates usually have heritabilities in the range of 0.3–0.8. The theoretical upper limit is 1 (i.e. the variance in the trait in the population is solely determined by additive genetic effects). Life-history traits of wild vertebrate populations usually have heritabilities below 0.4 (Gustafsson 1986). The same discrepancy also appears true for invertebrates (Houle 1992). This difference can partly be understood by the underlying evolutionary dynamics of selection acting on the trait which is known as Fisher's fundamental theorem of natural selection, after Sir Ronald Fisher who first described it (Fisher 1930). To understand the theorem we first need to deal with natural selection in a quantitative genetic framework.

Imagine a population that is subjected to a selection event and assume that we have measured the population before and after the selection event. The selection event could be survival or differential reproduction. The importance is that the selection events determines which individuals propagate their genetic material to the next generation. The selection differential is $S = \mu_1 - \mu$ where μ_1 is the population mean after the selection event and μ is the mean before (Fig. 2.3).

Table 2.2 Example of half-sib design to estimate heritability. as, among sires; ad, among dams; ap, among progeny (modified after Falconer and Mackay 1996).

Observational component	Convariance and estimated components	
Sires	$V_{as} = \text{Cov}_{(halfsibs)}$	$= 1/4 V_a$
Dams	$V_{ad} = \text{Cov}_{(fullsibs)} - \text{Cov}_{(halfsibs)}$	$= 1/4 V_a + 1/4 V_d + V_{ec}$
Progenies	$V_{ap} = V_p - \text{Cov}_{(fullsibs)}$	$= 1/2 V_a + 3/4 V_d + V_{ew}$
Total	$V_T = V_p$	$= V_a + V_d + V_{ec} + V_{ew}$
Sires + dams	$V_{as} + V_{ad} = \text{Cov}_{(fullsibs)}$	$= 1/2 V_a + 1/4 V_d + V_{ec}$
Three estimates of h^2		
$h^2_s = 4 V_{as}/V_{as} + V_{ad} + V_{ap}$	Contains no dominance effects	
$h^2_d = 4 V_{ad}/V_{as} + V_{ad} + V_{ap}$	Has dominance and maternal effects	
$h^2_{s+d} = 2(V_{as} + V_{ad})/V_{as} + V_{ad} + V_{ap}$	Has dominance and maternal effects	

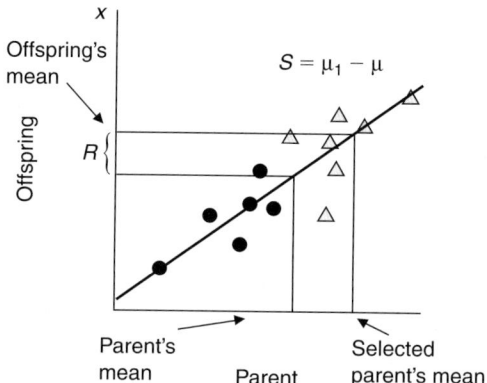

Figure 2.3 Schematic representation of the breeder's equation, the microevolutionary response to selection. S is the selection differential (the difference in mean values before μ and after μ_1 selection) and R is the response to selection in the next generation.

Now, to allow an evolutionary response to selection, a change in the mean values in the next generation, two things are needed. First, there needs to be additive genetic variance, V_A, in the population, which is the same as saying that we need significant h^2. Second, we need significant selection. The higher the values of h^2 and S the larger the response to selection. This can be formulated by the response to selection, R, being equal to the product of heritability and selection, such that

$$R = h^2 S$$

This the famous breeder's equation, which is illustrated in Fig. 2.3. There are numerous studies and reviews of natural selection and microevolutionary responses in the wild. Among the most famous examples are microevolutionary responses of beak lengths to varying selection pressures imposed by seed availability ultimately determined by weather conditions in Darwin's finches (Grant and Grant 1989).

The theory outlined above is a univariate case where epistatic and pleiotropic effects (genetic correlations) were ignored. When traits are correlated, the theory becomes somewhat more complex. Recall that $R = h^2 S$, which could also be written as

$$R = (V_A/V_P)S$$

When traits are genetically correlated the response to selection is described by

$$\Delta Z = \beta G$$

The bold typeface used here indicates we are working with matrices and G stands for the additive genetic variance–covariance matrix describing all the genetic variance and covariances (correlations) between any traits under consideration. β is the directional selection gradient describing selection on each of the traits. β can also be written as $\beta = S/P$ where S is a vector of the selection gradients on each of the traits and P is the phenotypic variance–covariance matrix. Thus

$$\Delta Z = (S/P)G = GP^{-1}S$$

In words, the response to selection equals the proportion of additive genetic variance (and covariance) multiplied by selection in accordance with the univariate case.

Fisher formulated his fundamental theorem in words by saying that 'the rate of increase in fitness of any organism at any time is equal to its genetic variance in fitness at that time' (Fisher 1930). That is to say, the evolutionary response depends on the heritability. In modern terms the theorem could be stated as follows: the rate of increase in the mean fitness of any organism at any time ascribable to natural selection acting through changes in gene frequencies is exactly equal to its genic variance in fitness at that time.

A formal proof of the theorem came from Maynard Smith (1998). Recall that $R = h^2 S$ and $S = \mu_1 - \mu$. The trait under consideration is fitness such that the weighted selection differential on fitness is

$$S = \Sigma k(\mu_1 - \mu)/N$$

where k is number of offspring to a parent and $N = \Sigma k$; that is, the total number of offspring. The number of offspring to a parent is that parent's fitness, W, and the weighted selection differential on fitness is

$$S = \Sigma W_i(W_i - \text{mean } W)/N$$
$$= \Sigma W_i^2/N - \Sigma(W/N) \text{ mean } W$$
$$= \text{mean } W_i^2 - (\text{mean } W)^2 = V_{\text{Pw}} \quad \text{(by the definition of variance)}$$

Substituted into the breeder's equation we get

$$R_\text{w} = h^2 V_{\text{Pw}} = (V_{\text{Aw}}/V_{\text{Pw}})V_{\text{Pw}} = V_{\text{Aw}}$$

which was to be proven.

By way of example we can understand the above argument by assuming that at a locus we have the following fitnesses for three genotypes: $w(AA) = 1$; $w(Aa) = 0.6$, and $w(aa) = 0.3$, and the starting frequency of A is very low, say 0.01. Thus the mean fitness in the population is close to 0.3. Now, when the A allele starts to rise in frequency owing to its selective advantage, the mean fitness in the population will rise accordingly as the allele frequency of A increases (Fig. 2.4). Seen over generations, both mean fitness and the allele frequency of A will increase in a sigmoidal manner and finally become fixed in the population. It can be seen that initially the heritability in the population is very low since initially almost all individuals have the aa genotype and thus there is very little additive genetic variation. Heritability increases in the population as the allele increases in frequency, reaching a peak when $p = q = 0.5$ and the mean fitness is 0.6. When the A allele starts to take over, the additive genetic variance starts to decline and accordingly so does the heritability.

The proof and the example above illustrate the important point that heritability is a dynamic property that changes with the evolutionary dynamics of the trait. Furthermore, it shows that traits closely related to fitness quickly become fixed and by necessity must have a heritability close 0. The closer the trait is related to fitness, the more likely the heritability is to be low. This is one explanation for why life-history traits have lower heritabilities than morphological traits. The life-history traits often measured in natural populations are those such as laying/weaning date, clutch/litter size, and longevity. Such traits explain more of the variance in fitness (lifetime reproductive success) than many morphological traits (Gustafsson 1988). There may of course be exceptions to this rule, when morphological traits explain much of the variance in fitness (e.g. Merilä and Sheldon 2000). However, in such circumstances such traits are predicted to have low heritabilities.

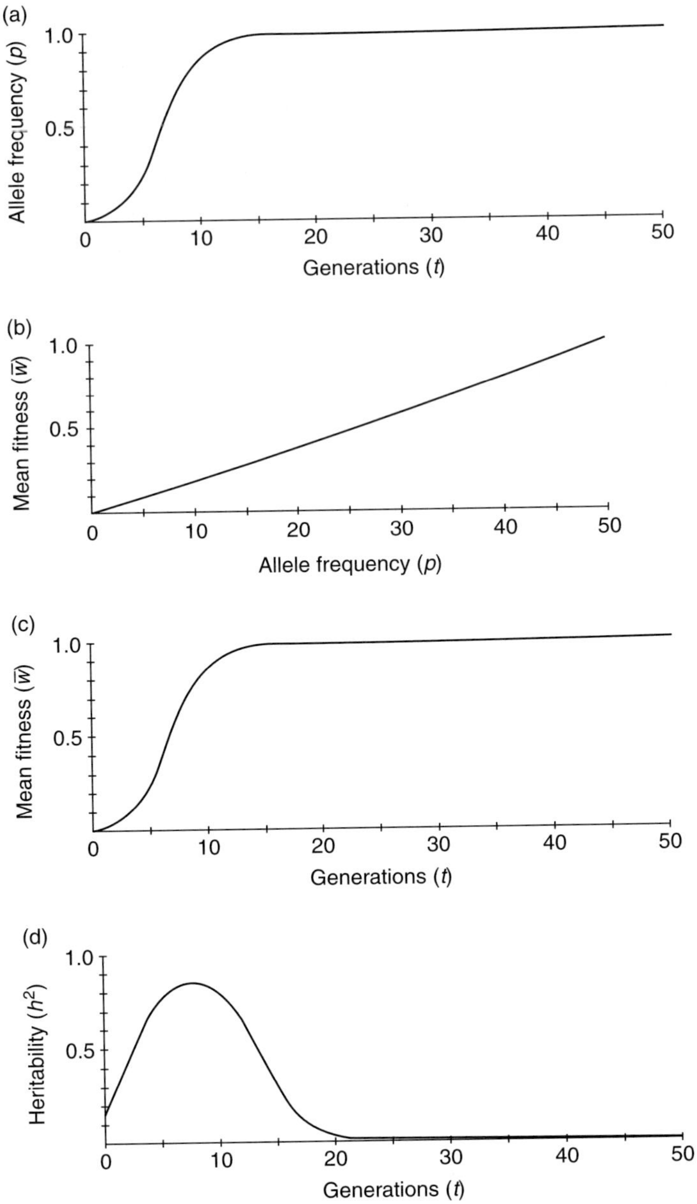

Figure 2.4 Schematic illustration of Fisher's fundamental theorem on natural selection. (a) Allele frequency change in response to selection. (b) The change in mean fitness in relation to allele frequencies. Change in mean fitness (c) and heritability (d) over time (generations). These graphs illustrate why traits related to fitness must have low or zero heritability.

An alternative explanation for low heritabilities of some traits is that traits differ in their susceptibility to environmental variance. Houle (1992) measured two morphological traits and two life-history traits in *Drosophila melanogaster*. In accordance with the theory outlined above the morphological traits measured, sternopleural bristle number and wing length, displayed relatively high heritabilities, 0.44 and 0.36, respectively, whereas two life-history traits, fecundity and longevity, showed low heritabilities, 0.06 and 0.11, respectively. However, Houle independently determined V_A for each of the traits and calculated a property that he termed evolvability. This is defined as $CV_A = 100(V_A)^{1/2}/P$ where P is the mean phenotypic value of the trait. Evolvability was high for bristle numbers, fecundity, and longevity (around 10) but low for wing length (1.56). This implies that the low heritabilities of the life-history traits is due to a high V_E and that given that this property is reduced, there should be ample additive genetic variance, V_A, for these traits to respond to selection.

As noted above, populations of endangered species are expected to show more subdivision than more numerous species. With genetic data it is possible to estimate population differentiation by calculating F_{ST}. We noted above that a general formulation of $F_{ST} = (H_T - H_S)/H_T$ (Nei 1975). Another way of formulating this is

$$F_{ST} = V_a/(V_a + V_b + V_w)$$

where V_a is the among-sample genetic variance component, V_b is the between-individual within-sample component, and V_w is the within-individual component (Weir and Cockerham 1984). Wright (1951) showed that for quantitative genetic data

$$Q_{ST} = V_{gb}/(V_{gb} + 2V_{gw})$$

where V_{gb} is the additive genetic variance among populations and V_{gw} is the additive genetic variance within populations. Thus population differentiation can be calculated also for quantitative genetic data (see also Chapter 6).

For practical purposes Q_{ST} can be obtained as

$$= (g)V_{pop}/(g)V_{pop} + 2(h^2)V_{err}$$

in which the variance components can be obtained from a standard one-way ANOVA. Here g is the assumed proportion of variance among populations due to additive effects, V_{pop} is the phenotypic variance due to populations, h^2 is the heritability of a trait within populations, and V_{err} is the phenotypic error variance (Lande 1992, Spitze 1993).

2.6 Conclusions

This chapter is a review of the most common techniques used in conservation genetics to study genetic variation. The list encompasses techniques at the phenotypic level from the basics of quantitative genetics, allozymes, various anonymous genetic markers such as AFLP and microsatellites, to DNA sequencing techniques. In Chapter 7 I will review genomic applications relevant for conservation studies.

3 Inbreeding, geographic subdivision, and gene flow

One of the major causes for deviation from the Hardy–Weinberg expectation is inbreeding. In the following I will outline the theory of inbreeding including a brief account on the theory of population subdivision and gene flow. This is of relevance to conservation issues because loss of habitat and fragmentation of habitats induces elevated levels of population structure in endangered species through reduced migration between remaining habitat fragments. Population structure is a major cause of inbreeding. Later in the chapter the relationship between genetic diversity and fitness will be discussed. In that context the issues of inbreeding depression and heterosis will be covered.

3.1 Inbreeding within populations

In understanding the processes that affect allele frequencies in natural populations and thus population structure, it is useful to start with the concept of an 'ideal population' (Wright 1931, 1938). An ideal population is a theoretical concept defined by Wright as 'the number of breeding individuals in an idealized population that would show the same amount of dispersion of allele frequencies under random genetic drift or the same amount of inbreeding as the population under consideration'. In an idealized population mating is assumed to be random and thus all parents have equal expectation of being parents of any progeny. It is sometimes argued that since no natural population ever mates at random, the ideal population has no real meaning. However, this misses the point as the ideal population defines the necessary standard for comparison. As we have already seen in Chapter 1, small population size leads to an accelerated loss of genetic variation, which is highly relevant in a conservation context. The effective population size N_e, defined as the population size that is expected given the observed allele frequencies assuming a randomly mating population (i.e. the idealized population size; Kimura and Crow 1963), is what matters from a conservation genetic perspective, not the census population size N.

The Hardy–Weinberg model is used to predict how allele frequencies in diploid populations are affected by mating and meiosis. During meiosis haploid gametes are produced by diploid individuals, which are then united during mating to form new diploid individuals. It is thus useful to be able to calculate genotypic frequencies from allelic frequencies and vice versa. This can be done using the well-known Punnet square. After one generation of random mating, any population will reach equilibrium where Mendelian segregation does not alter the allelic frequencies (Box 3.1).

As indicated above, non-random mating will cause deviation from the Hardy–Weinberg equilibrium. There are two such processes: assortative mating and inbreeding. Assortative mating, when similar individuals tend to mate with one another, will only affect the locus affecting mating and will change the homozygosity only at that locus. For example, if colour dimorphism is controlled by two alleles segregating at a locus, the heterozygosity at this locus will be lower than expected by chance if individuals of the same colour tend to mate with their own kind. Conversely, if there is disassortative mating (mating between divergent individuals being more likely) heterozygosity will increase at the loci affecting the trait. This may be the case with some major histocompatabilty (*Mhc*) loci where disassortative mating is sometimes observed (Milinski 2006). This may

---

Box 3.1 The Hardy–Weinberg model

In diploid organisms haploid gametes are formed during meiosis and mating produces new diploid individuals. It is therefore useful to be able calculate allele frequencies in gametes from knowledge of genotype frequencies in the zygotes (and vice versa).

This is done with the aid of the Punnet square. Given a diallelic locus, the frequency of AA is $P' = p^2$, that of Aa is $H' = pq + qp = 2pq$, and that of aa is $Q' = q^2$. At equilibrium,

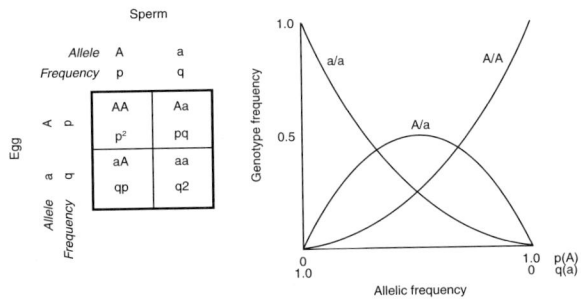

> Box 3.1 (Continued)
>
> $$p' = P' + H'/2$$
> $$= p^2 + 2(pq)/2$$
> $$= p^2 + pq$$
> $$= p(p+q) \quad \text{(and since } p+q=1\text{)}$$
> $$= p$$
>
> This shows that after one generation of random mating Mendelian segregation does not alter the allele frequencies.

explain why *Mhc* polymorphisms are prevalent in many populations (Piertney and Oliver 2006; see Chapter 5 in this volume).

Unlike assortative (or disassortative) mating, inbreeding will affect homozygosity on a genome-wide level. Inbreeding is defined as matings between individuals in a population that are more closely related than expected by chance. The inbreeding coefficient, f, describes the probability that two alleles at a locus are identical by descent. This is to say, both are copies of one particular allele inherited by common ancestry. The inbreeding coefficient is calculated by drawing two random alleles from the population. After the first allele is drawn there is a probability f that the second will be the same as the first. For example, if the probability of the first allele being A is p, then the probability that the second allele is A (identical by descent) is f. Thus, the probability that the two alleles are identical by descent (IBD) is

$$P_{IBD} = pf$$

The alternative to being identical by descent is identical by state (IBS); that is, the alleles are the same but they do not descend from the same ancestral state, which is given by

$$P_{IBS} = p^2(1-f)$$

where p^2 is the probability of drawing a similar allele and $(1-f)$ is the probability of not being identical by descent.

Thus the frequency of AA genotypes is

$$P = P_{IBD} + P_{IBS} = pf + p^2(1-f)$$

With some algebra and remembering that $p = 1 - q$ we get

$$P = p^2 + fpq$$

In general, it can be shown that

$$P = p^2 + fpq$$
$$H = 2pq - 2fpq$$
$$Q = q^2 + fpq$$

where the first terms after the equal signs are the usual Hardy–Weinberg expectations and the second terms describe the deviation from Hardy–Weinberg, as a consequence of inbreeding. It follows that if f tends to 1 (complete inbreeding), the heterozygosity, H, tends to 0 and thus inbreeding decreases the heterozygosity.

The inbreeding coefficient of an individual can be estimated from pedigrees (Wright 1969, Malécot 1948). Wright's method involves path analysis whereby, in a pedigree, the probability that an individual will be homozygous because of an ancestor shared on each side of the pedigree is calculated (Box 3.2).

In conservation studies of wild animals and plants, pedigree data are rare although some field studies have been able to infer pedigrees by observation or indirectly via genetic markers (Laikre *et al.* 1997, Kruuk *et al.* 2002). However, in captive populations, for example in zoos and botanical gardens, close records of the pedigrees are often kept and great care is taken to minimize the level of inbreeding by not mating close relatives.

Because pedigree data are so hard to collect in the field, researchers have often turned to indirect measures to estimate the level of inbreeding in wild populations. One obvious example of such an approach is to infer the level of inbreeding via observed deviations from Hardy–Weinberg expectations. In any population the level of inbreeding F is related to the level of heterozygosity, such as

$$F = (H_0 - H)/H_0$$

where H_0 is the heterozygosity expected from Hardy–Weinberg (the null hypothesis) and H is the observed level of heterozygosity in the population. This equation is often written as

$$F_{IS} = (H_e - H_o)/H_e$$

where H_e is the heterozygosity expected from Hardy–Weinberg and H_o is the observed level of heterozygosity (see below).

Box 3.2 **Pedigrees and path analysis**

We want to estimate Wright's inbreeding coefficient, f, the probability that two alleles at a given locus are identical by descent. Consider the following example pedigree. The sire Gustav has offspring with two dams: with Anna the son Erik and with Maja the daughter Pia. Pia and Erik have a son Kurt. What is the inbreeding coefficient of Kurt?

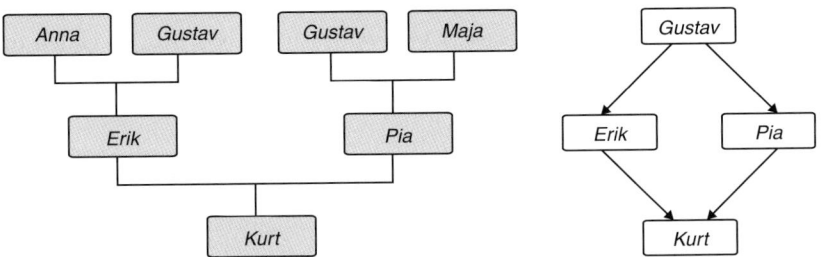

To simplify matters, we may start by only looking at the shared ancestors. Consider a gene of which Gustav has two different alleles, a1 and a2. Whichever is passed to Erik has a 50% chance of being passed to Kurt. At the same time, there is also a 50% chance that the same allele is passed from Gustav to Pia and a 50% chance it is passed from Pia to Kurt, if Pia has it. The total probability that Kurt will be homozygous for a1 or a2 because of the common grandfather is $0.5 \times 0.5 \times 0.5 = 0.125$ (12.5%).

Wright developed the method of path analysis to calculate inbreeding coefficients in pedigrees (Wright 1969). Applied to the example above the path from Kurt to the common ancestor Gustav and back again on the other side of the pedigree (Kurt–Erik–Gustav–Pia–Kurt) is determined, the number of individuals in the path excluding Kurt is counted (there are three: Erik, Gustav, Pia), and then 1/2 to the power of n (where n is the number of ancestors) is calculated. This gives

$$(1/2)^3 \text{ or } (1/2 \times 1/2 \times 1/2) = 1/8 = 12.5\% \text{ (as previously)}$$

Suppose there is one more generation, a great-grandfather of Kurt being the common ancestor. This adds one individual on each side of the pedigree:

$$f = (1/2)^5 = 1/32 = 3.125\%$$

> **Box 3.2** (*Continued*)
>
> In the case there is more than one common ancestor each closed path is counted and the probabilities are summed. As an example take the offspring of first cousins (who have two shared great-grandparents):
>
> $$f = (1/2)^5 + (1/2)^5 = 1/32 + 1/32 = 6.25\%.$$
>
> It is usually assumed that the common ancestor has $f = 0$. However, sometimes the f of the common ancestor is known. In such cases this f is added to the total probability. For example, if Gustav is the product of a first-cousin mating, Kurt's f value is:
>
> $$f \times (1 + f_A) = 0.125 \times 1.0625 = 0.133 = 13\%$$
>
> where f_A is the inbreeding coefficient of Gustav.
> In general:
>
> $$f = \sum_{i=1}^{N} (1/2)^{n-1} * [1 + f_A]$$
>
> where n is number of closed path lengths and N is the set of all common ancestors for those path lengths (Wright 1969).

In ongoing studies in my own research group of the locally critically endangered natterjack toad, *Bufo calamita*, in an archipelago off the west coast of Sweden we estimated F_{IS} from microsatellite loci in different island populations in the archipelago. We observed that allele frequencies deviated from Hardy–Weinberg expectations, and thus F_{IS} was positive in five populations and while standard errors were large in three, two populations were significantly different from what would be expected if there were no inbreeding in the populations (Rogell 2005). Three populations did not deviate from the null hypothesis of random mating (Fig. 3.1). Unfortunately, it would be premature to conclude that inbreeding is the cause of the elevated levels of F_{IS} observed (although this is a possible interpretation). A number of methodological concerns would first need to be eliminated to reach this conclusion. One alternative explanation for the observed patterns is that there may be so-called null alleles at some loci. Null alleles are alleles that do not amplify in the PCR reaction used to magnify and visualize the allelic variation at the loci under study. If such alleles are more prevalent in some populations than others, this may explain the observed increase in homozygosity and thus a wrongly inferred F_{IS} in those populations.

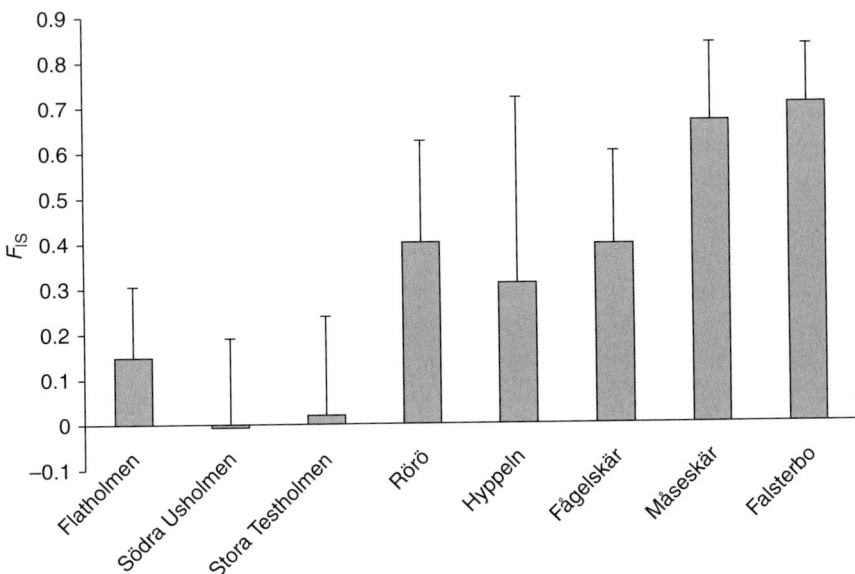

Figure 3.1 Mean F_{IS} (±1 SD) in subpopulations of natterjack toads on islands off the Swedish west coast (Rogell 2005).

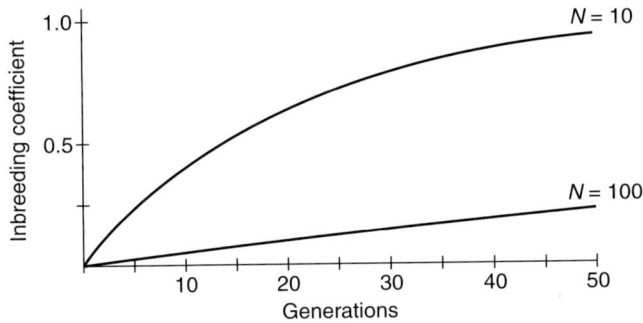

Figure 3.2 Inbreeding in relation to time (generations) in hypothetical closed populations of varying size (N is the population size). Inbreeding increases faster in small compared with large populations.

In theory, inbreeding is expected to increase in any closed population of finite size (Fig. 3.2). As noted above, the inbreeding coefficient is the probability that two alleles at a locus are identical by descent. This can also be interpreted as the probability that in the previous generation two alleles in two different individuals are identical by descent. $F = 0$ implies no inbreeding and $F = 1$ implies complete inbreeding and that all individuals are genetically similar. In a closed population (no immigration or emigration allowed) and if we ignore mutation (as mutation is an unlikely event and nearly non-existent in small populations), F increases over time. This is because genetic drift will cause the extinction of some alleles and fixation of others. The result is that individuals in later generations are more likely to carry alleles that are copies of the same ancestral alleles.

The rate at which inbreeding increases with time is

$$F = 1 - (1 - (1/2N))^t$$

where F is the inbreeding coefficient, N is population size, and t is the number of generations. Since genetic drift is the principal agent and is a stochastic process there is always random variation around the expected value. From the equation it can also be seen that inbreeding increases faster when N is small as compared to when N is large (see Fig. 3.2).

It follows that in closed populations, inbreeding increases over time even if the population is mating at random. However, real populations are often stratified into family groups of different sizes. In such stratified populations it has been suggested that the chance that individuals will mate more easily with a consanguineous individual (close relatives) is increased when population size is small (Cavalli-Sforza et al. 2004). This follows from the fact that there are a limited number of families and that families will share common ancestors more quickly when population size is small.

There are numerous examples of studies attempting to estimate the level of inbreeding from standing levels of genetic variation in natural populations. The following example is from my own research group's work on the black grouse *Tetrao tetrix* in populations in western Europe. The black grouse is a sedentary bird species adapted to the ecotone between open myre/moorland within taiga forest habitats. It has a breeding range from Britain in the west to the borders of China and Korea in the east. Not much is known about population trends in the east but the species has been carefully monitored in the western part of its range. Here, there has been a general decline in population size, range contraction, and fragmentation of habitats during the last 100 years (and perhaps longer; BirdLife International 2004). In some Western European countries where the species occurred in higher numbers previously, only small isolated remnant populations remain and in Denmark the species has gone extinct within the last few decades.

We sampled genetic variation at eight microsatellite loci in five populations and found that observed levels of heterozygosity did not depart from expected values in four of the populations (Table 3.1). Thus in none of these four populations was the inbreeding coefficient, F_{IS}, significantly different from 0 which would be the case if there is no inbreeding in these populations. Note that the estimate of F_{IS} may, for stochastic reasons, vary and become less than 0. If non-significantly different from 0, it should be interpreted as 0, meaning no inbreeding. However, in one of the populations, the one in northern England, we did detect a significant

Table 3.1 Genetic diversity in some European black grouse populations. The Category column refers to isolation and population size status, from large and continuous to small and isolated (from Höglund et al. 2007).

Category	Population	Year sampled	n	AR	H_e	H_o	F_{IS}
Continuous	Jyväskylä	1989–1995	57	4.49	0.74	0.66	**0.13**
	Østfold	1999	31	4.17	0.70	0.67	0.04
Contiguous	Abernethy	2000	16	3.93	0.63	0.65	−0.05
	Allgäu	1998–2000	23	4.38	0.73	0.69	0.06
	Ammer	1998–2000	18	4.20	0.73	0.71	0.02
	Vorarlberg	1998–2000	24	4.07	0.70	0.66	**0.06**
	Haut Savoi	1998–1999	9	4.16	0.72	0.67	0.07
	Tauern	1998–2000	27	4.12	0.70	0.69	0.01
	Tessin	1980–1983	16	4.61	0.73	0.64	**0.13**
Isolated	Northern Pennines	2000–2003	21	2.85	0.57	0.48	**0.15**
	Salland	2003	31	3.16	0.53	0.44	**0.17**
	Rhön	1992, 1995, 2003	8	3.93	0.72	0.55	**0.25**
	Waldviertel	2001–2003	14	3.27	0.56	0.57	0.01
	Llandegla	2004	8	2.81	0.52	0.53	0.00

AR, allelic richness (as per Goudet 2001) rarified to a constant sample size of 8; F_{IS}, inbreeding coefficient (bold indicates that the estimate is significantly different from 0); H_e, expected heterozygosity; H_o, mean observed heterozygosity; n, number of individuals analysed.

reduction in observed heterozygosity suggestive of past inbreeding (Höglund et al. 2007). Due to successful conservation efforts this population is now locally relatively abundant. However, like all English black grouse populations, the one sampled in the study has been severely threatened and its habitat fragmented during the last 100 years.

3.2 Population structure

As briefly discussed in the previous chapter, the extent of population subdivision is an important parameter in identifying and diagnosing threatened populations. When a large and widespread population is reduced in numbers it is likely to become locally extinct in areas where the previous population density was low for various reasons. If the population decline is severe, the range contraction following the decline may be so extensive that the emerging subpopulations may become relatively isolated. If left in isolation long enough, subpopulations evolve independently and local adaptation and genetic drift both contribute to building up genetic differences among subpopulations (Charlesworth et al. 2003).

46 *Inbreeding, geographic subdivision, and gene flow*

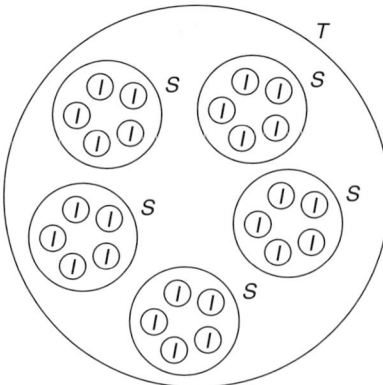

Figure 3.3 Schematic illustration of three levels of population structure: *T*, total population; *S*, subpopulations; *I*, individuals.

One consequence of a geographically subdivided population is that the structure causes inbreeding (Wright 1921, 1969). Imagine a large population that is split into many smaller populations of equal size. In such a system (metapopulation), gene flow between the subunits is the same as the probability that a random allele in any of the subpopulations is from a migrant, such that

$$m = \Delta p / (p_1 - p_2)$$

where Δp is the change in allele frequency after the migration event, p_1 is the allele frequency in the donor population, and p_2 is the allele frequency in the recipient population. Now gene flow, which is governed by the number of migrants, can be determined by multiplying m by population size N.

In the absence of selection, the genetic structure of any population is determined by both the inbreeding within the population and gene flow. The frequency and strength of these events determine the genetic structure. As with inbreeding within populations, population genetic substructuring can be assessed via deviations from Hardy–Weinberg expectations.

Assume that there are three levels of population structure. I is the level of individuals, S is the level of subpopulations, and T is the total population, for example the species under study (Fig. 3.3). It follows that H_I is the heterozygosity observed in an average individual, H_S is the average heterozygosity expected within randomly mating subpopulations, and H_T is the expected heterozygosity within the total population.

In a metapopulation the extent of population subdivision can be determined as

$$F_{ST} = (H_T - \text{mean } H_S) / H_T$$

This is a global fixation index estimating the extent of subdivision in the whole metapopulation. Wright (1969) showed that if one assumes equidistant subpopulations of equal size then

$$F_{ST} \approx 1/(1 + 4Nm)$$

It is sometimes useful to calculate pairwise F_{ST} values. In such cases all possible pairs of populations are considered separately. F_{ST} is calculated for each pair assuming that the two populations constitute the total population in each calculation. Calculated this way F_{ST} can be considered as an estimate of genetic distance among populations.

The reduction in heterozygosity due to local inbreeding within the subpopulations is given by

$$F_{IS} = (\text{mean } H_S - H_I)/\text{mean } H_S$$

This is a global measure; within any subpopulation F_{IS} is determined as shown in section 3.1.

The effect of both inbreeding and subdivision is

$$F_{IT} = (H_T - H_I)/H_T$$

It follows that

$$(1 - F_{IT}) = (1 - F_{IS})(1 - F_{ST})$$

As many threatened species are facing conservation problems that are related to fragmentation of previous ranges, smaller population sizes, and more and more isolated subpopulations it is evident that human impact has consequences for the patterns of inbreeding in threatened species. Thus we may predict that in general F_{IS} will tend to increase in small and isolated populations as a result of increased inbreeding. However, this is not always the case and I will return to possible causes for why not below. Another general prediction relevant for conservation is that F_{ST} among subpopulations tends to increase as a consequence of population fragmentation.

3.3 Effective population size

In almost any application in conservation biology population size is one of the most important parameters to understand. As already noted in Chapter 1 and above, the population size that matters in conservation genetic studies is not always the census population size. Instead it is the number of individuals which actually reproduce and propagate their genetic material to future generations that is the determining factor for future genetic variation. Therefore geneticists are concerned with the effective population size, N_e. Theoretically this is defined as the population size that is expected given the observed allele frequencies assuming a randomly mating population.

With knowledge of the mutation rate μ it is possible to calculate the population parameter θ as

$$\theta = 4N_e\mu$$

Thus with knowledge of the heterozygosity in the population and the mutation rate it is possible to determine effective population size, for example via

$$H_e = 1/(1 + \theta)$$

N_e depends on many factors but the three most important are as follows.

1. Variation in reproductive success (sometimes referred to as variation in family size). Here N_e is given by

$$N_e = 4N/(\sigma^2 + 2)$$

where σ^2 is the variance in family size and N is the actual population size. Thus, for example in some marine fishes, variation in reproductive success can be extreme. Some parents may give rise to thousands of offspring in a single reproductive event, whereas others fail completely. N_e can thus be quite low despite large stocks and census population sizes.

2. Unequal sex ratios. When there are an unequal number of reproductively active males and females, N_e is given by

$$N_e = 4N_m N_f / N_m + N_f$$

where N_m and N_f are the effective number of males and females, respectively. N_e is always highest when there is a 50:50 sex ratio, dropping off the more the sex ratio becomes skewed.

3. Population size fluctuations will affect N_e. More precisely the effective population size is decided by the harmonic mean:

$$N_e = n/((1/N_1) + (1/N_2) + \ldots (1/N_n))$$

where n is sample size and $N_1, N_2 \ldots N_n$ are temporal population size estimates. Here it can be seen that even one small population-size estimate will heavily influence the mean value and hence N_e.

As indicated above it is at least theoretically possible to calculate N_e in a Wrightian population; that is, a population defined as being a unit in Hardy–Weinberg equilibrium. However, there is considerable difficulties in applying effective population size to real populations (Waples and Gaggiotti 2006). In ecological theory populations are sometimes assumed to persist in a balance between extinction and colonization. In such populations of populations, or so-called meta populations, the theoretical population size is unknown but attempts have been made to reach generalizations via simulation (see papers in Goldstein and Schlötterer 1999).

Estimating N_e in wild populations with overlapping generations is not a trivial task (Jorde and Ryman 1995). N_e can be estimated both globally and locally and in the long and short term. The N_e estimated from heterozygosity is the long-term N_e. Jorde and Ryman (1995, 1996) introduced a method based on observed temporal shifts in sample allele frequencies. This method have recently been modified to account for biases due to small sample sizes and when allele frequencies are highly skewed (Jorde and Ryman 2007). In small populations with a few breeders allelic frequencies can change rapidly and thus indirect point estimates of N_e may be dependent of what particular year class was measured. Likewise, population differentiation

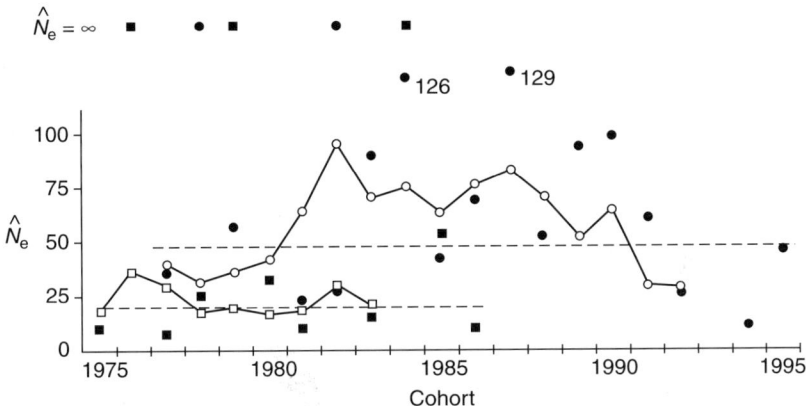

Figure 3.4 Point estimates of effective size for pairs of consecutive cohorts of brown trout (and for two localities, respectively), and the corresponding (harmonic) mean N_e (open symbols) obtained from moving averages for F over five consecutive cohorts (i.e. four cohort pairs). Cohort on the x axis represents the first cohort used for each N_e estimate. Dashed lines indicate the total estimate for each population. Note the broken y axis and that large N_e point estimates are given as numbers (∞ = infinity) (from Palm et al. 2003b, reprinted with permission from the publisher).

estimates may be biased. One way of increasing the precision of the estimates is to use temporal genetic data for estimating N_e and study the temporal stability of population structure. Palm and coworkers (2003b), using Jorde and Ryman's method, showed that individual point estimates of N_e may vary considerably between years. The authors assessed the amount of spatiotemporal genetic variation at 17 allozyme loci and estimated current N_e in two populations of stream-resident brown trout, *Salmo trutta*, using data collected over 20 years (Fig. 3.4).

3.4 Examples of population structure in endangered species

In my group's studies of the natterjack toad in the Bohuslän archipelago on the west coast of Sweden we observed substantial population differentiation among islands, when estimated both with microsatellite loci ($F_{ST} = 0.25$, $P < 0.0001$; Rogell 2005) and with amplified fragment length polymorphism (AFLP; $F_{ST} = 0.13$, $P < 0.001$; Thörngren 2006). This suggests that migration between these populations has been limited in the past and that any possible colonization/recolonization events may have been subjected to strong founder events mediated by the population bottlenecks induced by a few colonists.

Likewise, we observed strong genetic differentiation among the three remaining and fragmented distributional areas of black grouse in Britain ($F_{ST} = 0.13 \pm 0.04$, $P < 0.001$; see Box Fig. 4.1; J.K Larsson *et al.*, unpublished results). British black grouse are known historically to have had a wider and not so fragmented distribution. It is highly unlikely that there is any present-day migration between the remaining metapopulations and thus each of these are evolving independently of one another and should be treated as separate management units.

As mentioned before, it is important to determine population structure in endangered species, because fragmented populations are expected to become differentiated if migration between remaining units becomes impaired. Since population densities often are reduced at the limits of species' distributions, populations become more fragmented (Hampe and Petit 2005). More differentiation, as measured by F_{ST}, is thus predicted at species margins. This is a prediction which seems to be often, but not always, borne out (see Lönn and Prentice 2002, Mandak *et al.* 2005, Höglund *et al.* 2008).

Studies of both capercaillie *Tetrao urogallus* and black grouse suggest that gene flow is impaired in marginal habitats at the range of the distribution. In capercaillie pairwise F_{ST} estimates become larger with increasing distance in the northern range of the Alps, a pattern consistent with an hypothesis of isolation by distance (Segelbacher and Storch 2002). In the central Alps connectivity among populations is higher and the pattern of isolation by distance is not prevalent. Similarly, in black grouse F_{ST} and isolation by distance was stronger in the

French Alps, at the south-western margin of the species' distribution, than at more central areas. In Finland, which is more at the core of the distribution, there is more connectivity among the populations (Caizergues *et al.* 2003).

The perennial outcrossing plant *Gypsophila fastigiata* grows in a patchy distribution on the island of Öland in the Baltic Sea. It was shown that gene diversity in allozyme loci was lower in peripheral populations than populations more centrally located in the network (Lönn and Prentice 2002). The authors explained the lower diversity in peripheral populations by a combination of genetic drift (more drift in smaller populations) and lower levels of gene flow (lower in more isolated populations).

3.5 Inbreeding depression

Inbreeding may become prevalent, especially in small and isolated populations. Inbreeding is manifested through non-random mating and, as concluded above, reduces heterozygosity. It follows that the opposite of inbreeding—outbreeding—may increase the level of heterozygosity. Neither inbreeding nor outbreeding as such may have any fitness consequences and thus need not be harmful to populations. However, when there are negative fitness effects on individual phenotypes, inbreeding becomes of particular concern to conservation biology.

Under certain circumstances inbreeding may lead to inbreeding depression and generally outbreeding leads to so-called heterosis (hybrid vigour). If the mating individuals are too genetically dissimilar, however, outbreeding may lead to negative fitness effects, known as outbreeding depression. It follows that there is an optimal level on the inbreeding–outbreeding continuum.

There are two general, and not necessarily exclusive, hypotheses of why inbreeding may lead to inbreeding depression (Charlesworth and Charlesworth 1987, 1999). The first, the so-called partial dominance hypothesis, states that inbreeding depression is due to the effects of recessive deleterious alleles. Recessive lethal or nearly lethal alleles segregate in many populations at low frequency. When inbreeding increases homozygosity, the chance that any of these alleles will be found in the homozygous state, and thus expressed at any locus, is increased. The second explanation, the overdominance hypothesis, states that inbreeding depression is caused by a general decline in heterozygosity in inbred populations. It has sometimes been observed that heterozygous genotypes have a superior performance (heterozygote advantage) over any homozygous genotype (e.g. the famous case of sickle cell anaemia and resistance to malaria in humans). With inbreeding there is a general genome-wide reduction in heterozygosity which may cause a general decline in overdominance and thus cause inbreeding depression.

Under both of these hypotheses the extent of inbreeding depression in a population depends on the genetic load of the population. Genetic load is defined as the accumulation of recessive alleles and/or loss of heterozygote advantage. However, there is one important difference between the two hypotheses. Under partial dominance natural selection will eventually remove the alleles causing inbreeding depression. This cannot happen with overdominance.

In recent years, it is fair to say that the partial dominance hypothesis has received more attention although researchers are always careful to point out that both processes may occur simultaneously. In the few species in which inbreeding depression has been studied carefully about half of the effects of inbreeding are due to recessive lethal alleles and the rest due to loss of heterozygote advantage (or other genetic mechanisms that are not diminished by natural selection; Lacy and Ballou 1998).

The number of lethal equivalents per diploid genome is an estimate of the average number of alleles per individual in the population if all deleterious effects of inbreeding were due entirely to the expression of recessive lethal alleles (Morton *et al.* 1956). This means that in a population in which inbreeding depression is prevalent, one lethal equivalent per diploid genome may mean one recessive lethal allele per individual, or there may be some other combination of recessive deleterious alleles which equates to this in effect.

One method to estimate inbreeding depression is via the logarithmic model:

$$\ln(S) = A - Bf$$

where S is survival (or some other fitness measure), f is the inbreeding coefficient, and A and B are parameters. Thus in a pedigree or in experimental crosses the inbreeding coefficient of each individual in a sample is determined and regressed against the logarithm of survival (Morton *et al.* 1956). A may thus be interpreted as the logarithm of survival in the absence of inbreeding and B is the portion of the lethal equivalents per haploid genome. Recent results using this approach relevant to conservation has been reported, for example, in the guppy (Nakadate *et al.* 2003, van Oesterhout *et al.* 2007).

Inbreeding depression is prevalent in captive and experimental populations and is variable in extent both among and between species and study populations (Lacy and Ballou 1998). Studies of semi-captive populations have shown that inbreeding depression becomes more severe under less benign conditions. For example, in studies of mice inbreeding depression was considerably more pronounced when the mice were living in semi-captive conditions as compared to the more benign laboratory environment (Meagher *et al.* 2000). In a review including data from seven bird species, nine mammal species, four species of poikilotherms (snakes, fish, and snails), and 15 plant species Crnokrak and Roff (1999) found 169 estimates of inbreeding depression for 137 traits. They found that inbreeding was more severe under natural conditions as compared with presumably more benign conditions in captivity. In small populations, characteristic of many endangered species, all individuals may suffer from inbreeding depression because of the cumulative effects of genetic drift that decrease the fitness of all individuals in the population (Hedrick and Kalinowski 2000).

In studies of wild animals, island populations have long been used for long-term studies, mainly because on islands the geographical limits of the study population are made more easily. Such long-term studies of marked individuals have revealed inbreeding depression in great tit *Parus major* (van Nordwijk and Scharloo 1981), song sparrow *Melospiza melodia* (Keller 1998), two species of Darwin's finches (Gibbs and Grant 1989, Grant and Grant 1995), and collared flycatchers *Ficedula albicollis* (Kruuk *et al.* 2002). Similarly, a study of an island population of red deer *Cervus elaphus* has shown inbreeding depression in the wild (Coulson et al 1998, 1999, Slate *et al.* 2000).

A small, introduced population of muskoxen, *Ovibus moscatus*, resides in the Norwegian mountains on the border to Sweden. Five animals immigrated to Sweden in 1971 and inbreeding depression has been inferred in the Swedish population, which is very likely to go extinct in the near future (Laikre *et al.* 1997; Fig. 3.5).

Despite the overwhelming support for the prevalence of both inbreeding and inbreeding depression in natural populations there are still studies that fail to detect inbreeding depression in studies of endangered species. Many plant species and populations of plants are self-fertilizing. This means that at least sometimes, if not always, inbreeding is complete ($F = 1$) in such populations (Schemske and Lande 1985). In animals, some populations with known severe inbreeding show no detectable signs of inbreeding depression (e.g. Groombridge *et al.* 2000). One explanation for the absence of inbreeding depression in these cases is that the population history may affect the severity of inbreeding depression. This may also explain the observation that in captive and experimental populations, inbreeding depression is variable in extent both among and between species and study populations.

During inbreeding or during population-size bottlenecks, genetic variation is lost and along with a general loss of genetic variation the deleterious variation is also lost. In other words, the genetic load of the population may become reduced. However, while purging of deleterious recessives by natural selection may occur under some circumstances both theoretical and empirical evidence question the effect of population-size bottlenecks. Thus a distinction between slow and fast inbreeding is often made.

Under slow inbreeding natural selection is allowed to act upon a population for many generations (Frankham *et al.* 2001). In a theoretical study (Kirkpatrick and Jarne 2000) showed that inbreeding depression decreases immediately after a sudden reduction of population size, but the drop is modest even for severe bottlenecks. Highly recessive mutations experience a purging process that causes inbreeding depression to decline for a number of additional generations but the

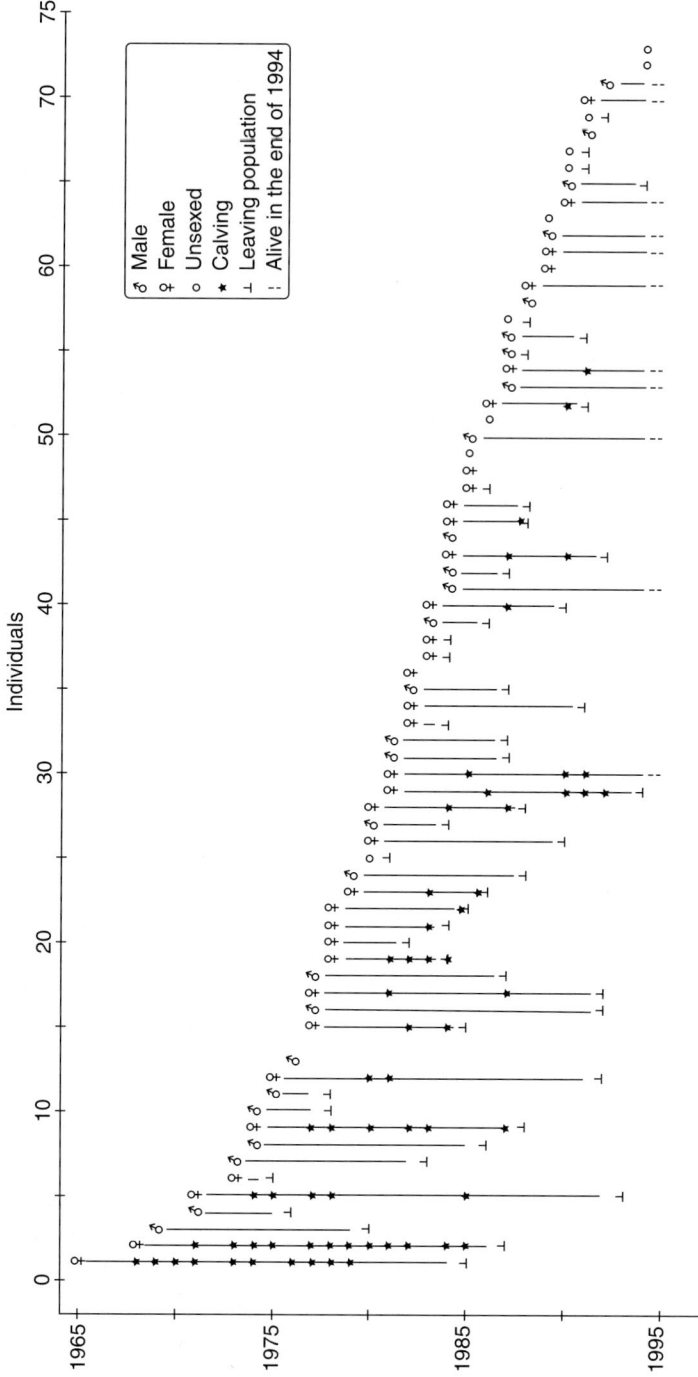

Figure 3.5 Lifespan in an inbreed herd of muskox in Sweden. The herd was followed from its emigration from Norway to Sweden and there was no immigration into the herd after it was founded (from Laikre et al. 1997, reprinted with permission from the publisher).

absolute fall in inbreeding depression may often be only a few percentage points for bottlenecks of 10 or more individuals.

It has thus been suggested that in captive populations the breeding programme ought to mimic slow inbreeding as much as possible. However, there is controversy regarding the effectiveness of purging in reducing the extinction risk. Frankham and coworkers (2001) evaluated the effects of purging on the extinction risk due to inbreeding in experimental *Drosophila melanogaster* populations. Overall there were small and non-significant differences between the extinction rates in the non-purged and purged treatments, indicating that the effects of purging were small.

Under fast inbreeding the period allowed for natural selection to act is too short to remove lethal alleles from the population and thus fast inbreeding is not predicted to have any measurable effect on inbreeding depression. Observe that the terms fast and slow inbreeding refer to extremes of a continuum and that here is no categorical difference between the two.

3.6 Heterozygosity–fitness correlations

As outlined above, inbreeding may severely hamper individual survival and performance and thus contribute to population declines and eventually local extinction. Much focus in conservation biology has therefore been directed towards detecting and measuring the negative effects of inbreeding in endangered populations (review in Hedrick and Kalinowski 2000, Keller *et al.* 2006). Ideally inbreeding can be estimated with the aid of pedigree information (Wright 1922), a method commonly employed to minimize inbreeding within zoo populations (Kalinowski and Hedrick 1998). However, pedigree information is not easily obtained in wild, free-ranging populations. Researchers have therefore tried various methods to estimate the negative effects of inbreeding indirectly. Most of the methods are based on the logic that inbreeding reduces heterozygosity and therefore less-heterozygous individuals should be more inbred (Coltman *et al.* 1998, 1999, Coulson *et al.* 1998, Pemberton *et al.* 1999). This approach has lately been criticized severely on both empirical and theoretical grounds and methods to infer pedigrees via molecular data are strongly advised (Pemberton 2008). Nevertheless, many studies have indeed found statistical associations between various measures of individual heterozygosity and measures of individual performance (e.g. Slate *et al.* 2000, Rossiter *et al.* 2001; see Kempenaers 2007 for a review).

One such study has been performed by my own research group. We used data from a large sample of male black grouse whose performance had been monitored in the field and which had been genotyped at 15 microsatellite loci. Male lifetime lekking performance was studied, and related to indirect measures of inbreeding in a wild population in central Finland between 1989 and 1995 (Höglund *et al.* 2002). Inbreeding was approximated with two estimates of heterozygosity (the lower the heterozygosity the greater the inbreeding). We found a significantly positive relationship between one of the measures of heterozygosity and lifetime copulation success (LCS), while the relationship of the other heterozygosity measure with LCS was close to significant. We also found that males that never obtained a lek territory had lower mean heterozygosity than males that were observed on a territory during at least one mating season in their life. Furthermore, among males that were successful in obtaining a lek territory, LCS and heterozygosity were highest for those males that held central territories. These data imply that heterozygous males had an advantage in the competition for territories. Whether these correlations are ultimately driven by inbreeding is unknown but if they are, inbred males have lower fitness than outbred males.

This study was one of the first attempts to link measures of inbreeding and lifetime fitness in a non-isolated population. It is important in establishing that the relationships found in previous studies on closed islands and in captive populations are not artefacts of low gene flow created by limited dispersal but a general feature of wild vertebrate populations. Furthermore, if signs of inbreeding depression could be found in the large, connected, and numerous Finnish population, then this suggests that inbreeding effects should not be ignored in the conservation of black grouse. It remains to be shown under what circumstances inbreeding has a negative effect. It may be predicted that populations that have undergone rapid fragmentation and contraction in numbers (fast inbreeding) should suffer more from inbreeding depression than populations that have been subjected to a sustained period of population reduction (slow inbreeding). Thus the negative effects of inbreeding could be even stronger in threatened and isolated black grouse populations in central and Western Europe than in Finland if subjected to fast inbreeding. Alternatively, these populations may have been purged if subjected to slow inbreeding.

While there is ample evidence from a wide range of organisms of a relationship between heterozygosity and fitness (Kempenaers 2007) it is less clear that the relationship between heterozygosity and inbreeding is as straightforward. The following quote is from Kempenaers (2007): 'there is much scepticism about heterozygosity-fitness correlations. The generality and magnitude of the effect have been repeatedly questioned, and there is an ongoing debate about whether the correlation reflects inbreeding or something else.'

The generality of the effect has been questioned on two grounds. First, there may be a publication bias in favour of studies that do find a significant correlation while studies that do not find the relationship fail to become published. Second, the effect sizes reported are often small (Coltman and Slate 2003). It has been pointed out that since the effect is generally small, a large number of individuals and marker loci need to be involved to avoid statistical Type II errors and something in the order of 10 000 genotypes (individuals multiplied by the number of loci) need to be studied to allow meaningful interpretations (Slate and Pemberton 2002). These are numbers rarely, if ever, reached in empirical studies, especially in endangered and small populations.

Part of the debate on how to interpret heterozygosity–fitness correlations concerns the mechanism involved in generating a positive correlation, if it exists. Three hypothesis have been put forward (see Hansson and Westerberg 2002 for a review). The first hypothesis is the so-called general- (or global-) effect hypothesis. Under this hypothesis, a positive correlation between heterozygosity and fitness is driven by the genome-wide loss of heterozygosity due to inbreeding and the negative effects of inbreeding depression. The marker loci used are, under this hypothesis, not directly involved or linked to loci causing inbreeding depression, but are selectively neutral markers of a genome-wide loss of heterozygosity. The hypothesis thus predicts that the heterozygosity at all marker loci used should be correlated.

The local-effect hypothesis suggests that the correlation between heterozygosity and fitness is due to the negative effects of homozygosity at functional loci. Thus the effect is driven by linkage disequilibrium between particular marker loci and particular loci affecting fitness. This hypothesis thus predicts that the heterozygosity at the marker loci should be uncorrelated.

The final hypothesis is the direct-effect hypothesis that explains the heterozygosity fitness correlation by a direct effect of particular marker loci. This effect is believed to be most severe when the marker loci used are allozymes or functional genes (like *Mhc* loci). This hypothesis is thus not assumed be applicable to microsatellites which are almost invariably assumed to be neutral. However, there is increasing evidence that microsatellite repeat numbers may sometimes have functional significance by, for example, influencing replication and gene expression (e.g. Chistiakow *et al.* 2006). However, for most microsatellites the direct effect is most likely of minor importance.

There seems to be a general consensus that the general-effect hypothesis cannot explain all heterozygosity–fitness relationships. Thus inbreeding, measured by inbreeding coefficients, as a single and general explanation to these relationships, is refuted (Coulson *et al.* 1998, Balloux *et al.* 2004, Pemberton 2004, Slate *et al.* 2004). Instead two more complicated scenarios are envisaged. First, it has been proposed that inbreeding coefficients do not completely estimate the total proportion of an individual's alleles that are identical by descent (Markert *et al.* 2004). To see this: full sibs on average share 50% of their genomes and have an inbreeding coefficient of 0.25. This is on average: individual dyads could still share more (or less) of their genomes. This explanation thus implies that multilocus heterozygosity is a better measure of inbreeding and susceptibility to inbreeding depression than inbreeding coefficients and could explain why heterozygosity–fitness correlations are found even within groups of individuals with the same inbreeding coefficient.

The other explanation for heterozygosity–fitness correlations is the local-effect hypothesis, that some marker loci are in physical linkage disequilibrium with selected parts of the genome. This explanation appears to be relevant in some but not all empirical studies (reviewed by Kempenaers 2007; see also Ferreira and Amos 2006). As far as current evidence goes it seems prudent to suggest that both of these explanations could be considered when trying to understand empirical data.

3.7 Rescue effects

Ecological theory predicts that immigrants from surrounding populations can prevent the extinction of small populations. Brown and Kodrick-Brown (1977) mentioned two reasons for the rescue effect. One is the demographic boost of the endangered population, a process known as the demographic rescue effect. Another suggested reason for why migration might rescue populations is that immigrants may increase the genetic variation in the population. This would reduce inbreeding depression and increase the adaptive potential. This particular effect has been termed genetic rescue (Ingvarsson 2001).

In several studies (some already mentioned in this book; see Chapter 1) the authors have suggested the discovery of a genetic rescue effect. The rescue has either occurred by natural immigration (in the case of Scandinavian wolves: Ingvarsson 2002, Vilà *et al.* 2003, Liberg *et al.* 2005) or by intervention by conservationists moving animals from elsewhere into threatened populations (in the case of Florida panthers, Pimm *et al.* 2006a; Illinois prairie chicken, Westemeier *et al.* 1998; Swedish adders, Madsen *et al.* 1999, 2004). In each of these examples the claims have been made that a previously dwindling population, each with signs of inbreeding depression for traits related to fecundity or aberrant traits indicative of inbreeding, have disappeared after the introduction of new blood. In each of these cases there has also been evidence of populations which previously exhibited negative growth reversing this trend and starting to grow in size.

None of these examples is uncontroversial (see for example Creel 2006, Maehr *et al.* 2006, Mills 2006, Pimm *et al.* 2006b, Culver *et al.* 2008 in the case of the Florida panther). All of the studies involved monitoring of free-living populations subjected to various conservation efforts. As such they are not controlled laboratory experiments where potential confounding factors except genetic ones can be controlled for. For example, as well as moving animals from genetically more diverse populations to a threatened one there has typically also been other measures taken to improve the conditions of the focal population, such as habitat improvements, predator control, supplemental feeding, etc. Thus it may be difficult to ascribe any improvement to the genetic effect even if the results are as predicted and there had been a genetic restoration. The explanations to an increase could be: (1) an increase in genetic variation which may release the population from adverse genetic effects such as inbreeding depression, (2) demographic effects derived by the increase of population size, or (3) a combination of the two.

We examined the effects of a supported release of green toads *Bufo viridis* into a critically endangered population on the small Baltic island Utklippan (B. Rogell *et al.*, unpublished results). This supported release resulted in a rapid increase in population size. With AFLPs we estimated the genetic variability in both the post-introduction Utklippan population and in the supported release population. The allele frequencies in the two populations were used to calculate which effective population size that would result in the observed amount of genetic drift over one generation and we were able to show that the recovery after the supported release was associated with a very strong bottleneck (N_e was less than two individuals). Therefore, it is unlikely that the successful supported release can be attributed solely to a genetic restoration, and that demographic effects are likely to be highly important in this case. However, neutral genetic variation may be less informative than quantitative measurements of variability (Reed and Frankham 2001). It is therefore not possible to completely exclude the importance of genetic restoration of the green toad population on Utklippan.

3.8 Conclusions

Inbreeding is a fact in any closed, non-randomly mating population and inbreeding becomes more severe when population size is small. If inbreeding leads to inbreeding depression this may have severe consequences for threatened populations. Thus, several authors have advocated so-called genetic rescue projects in which the genetic variability of natural populations may be restored by transplantation from other populations of the same species. However, such genetic rescue projects are not uncontroversial and in most of the published cases there has been a debate as to whether any possible positive effect is due to the restoration of the genetic diversity or simply due to a demographic effect. It seems safe to conclude that when effective population size has become low ($N_e < 10$) in a short space of time and when there are clear signs of inbreeding depression in morphological, physiological, and life-history traits, genetic rescue of an endangered population should be considered.

4 *Genetic diversity in changing environments*

Throughout the history of life on Earth environments have been changing and taxa that have been unable to adapt to these changes have become extinct (Erwin 2006). At the time of writing, perhaps the most discussed topic in the natural sciences and in society at large are fears of the effects of human-induced changes on the environment and in particular the effects of global warming (IPCC 2007). The climate has changed before, and in the northern hemisphere climates were considerably colder some 20 000 years ago during the Pleistocene ice ages. It is clear that some species in the northern hemisphere (e.g. reindeer and arctic fox) are threatened because at the present stage of the glaciation cycle their habitats are diminishing (Dalén *et al.* 2005). This process will be speeded up by the present human-induced climate change, which is making the climate even warmer. Thus some cold-adapted species live in shrinking environments and are retreating from their original ranges. The patterns of genetic variation in such species can be contrasted to what is known from species that are expanding their range because of a warmer climate. Species that are expanding are interesting because they may produce insights into what makes a successful colonist. Such studies are clearly linked to studies of introduced and invading species which have become a major threat to the feral species around the world. What ecological and genetic constitutions are required to become a successful invader?

4.1 Fragmentation and natural and human-induced barriers to gene flow

As argued before in this book, population fragmentation and isolation may have extremely detrimental effects on the fitness and viability of extant populations, and also the evolutionary potential of species (see papers in Ferrière *et al.* 2004). It is therefore important to understand what is causing fragmentation and how to ameliorate its effects. A general effect of the growth of the human population is that natural habitats are lost. The remaining habitat is further cut up into smaller and smaller pieces with increasing distance among remaining habitat fragments. Also, human infrastructure such as roads, railways, and other constructions may reduce the movement of organisms and impose barriers to migration. The minimum viable population size is often not maintained in anthropogenically isolated populations occupying fragmented habitats. The consequence of this induced population genetic structure is predicted to affect population persistence and long-term survival where small and isolated populations run a higher risk of extinction (Frankham et al. 2002, Goossens et al. 2006).

Under the extinction vortex scenario, small and isolated populations are subjected to increased levels of inbreeding due to reduced migration and increased genetic drift in a downward spiral towards extinction (Loeschke et al. 1994). As has been argued previously, lost genetic variation may furthermore affect any population's ability to adapt to future changing selection pressures (Soulé 1976, Lande 1988, Frankham 1996). However, the theoretical consequences of habitat fragmentation on population viability have only rarely been tested in natural populations at appropriate spatiotemporal scales (Hitchings and Beebee 1998, Landweber and Dobson 1999). This hampers our ability to manage natural populations on the basis of genetic data and to ameliorate the effects of habitat alteration.

One reason why progress in the empirical study of human-induced fragmentation has been slow is perhaps that early analytical tools, such as F_{ST} analyses, relied on populations being defined *a priori*. In some circumstances and when there are clear geographic breaks in the distribution of a species, such structure may be possible and easy to infer. However, natural populations are often not clearly definable in this way. Therefore, a number of statistical tools have been developed that do not require a prior definition of population structure, but instead allow researchers to define population structure from their data.

The perhaps simplest way to detect whether there is any population structure in a sample of genotyped individuals would be to produce a two-dimensional plot of the genetic structure. If there are separate clusters in such a plot, population structure may be inferred. There are several techniques, all akin to principal component analyses, that reduce multilocus variation to two dimensions (Box 4.1).

Multivariate analyses such as principal component analysis (PCA) or multi-dimensional scaling (MDS) may thus help in identifying clusters of populations and individuals. However, a caveat is that these graphical methods are only indirectly connected to statistical procedures which allow identification of homogeneous clusters of individuals (Evanno et al. 2005).

In the following are a few examples of how geographic structures may be identified from multilocus genetic variation. In European capercaillie, *Tetrao urogallus*, there is a clear geographic structure along the first principal component

Box 4.1 Techniques for visualizing multilocus genetic data in two dimensions

All these techniques can be found in most standard statistical software. When implemented in widely used population genetic software, a reference is given to the particular software.

In general, these ordination techniques have two purposes. The first is to reduce the number of variables in an analysis (if applied to two-dimensional plotting, this would mean two variables). The second purpose is to classify variables in groups describing related aspects of the studied variation. For example, in the case of morphology variables may be classified as belonging to say size and shape. In population genetics these techniques are most often used to display multilocus genetic data or genetic distances between populations in two dimensions (Legendre and Legendre 1988, Quinn and Keough 2002).

As shown in the examples these analyses can be used to visualize genetic distances between populations as population means (Box Fig. 4.1a; PCA, data from European capercaillie in Segelbacher et al. 2003), genetic distances between populations and individuals (Box Fig. 4.1b; CA, data from British black grouse, J.K. Larsson et al., unpublished results; Box Fig. 4.1c; PCO, data from central European black grouse in Larsson et al. 2008), and genetic distances between populations (Box Fig. 4.1d; MDS, data from turbot in Florin and Höglund 2007). Note that the choice of data in these examples is arbitrary; in principle any of the techniques could have been used in all of the examples.

Name	Abbreviation	Software	Reference	Internet address
Multidimensional scaling	MDS			
Principal coordinates analysis	PCO	GenAlEx	Peakall and Smouse 2006	http://www.anu.edu.au/BoZo/GenAlEx/
Principal component analysis	PCA	PCAGEN		http://www2.unil.ch/popgen/softwares/pcagen.htm
Correspondance analysis	CA	Genetix	Belkhir et al. 1996–2001	http://www.genetix.univ montp2.fr/genetix/genetix.htm

Fragmentation and barriers to gene flow 63

Box 4.1 (*Continued*)

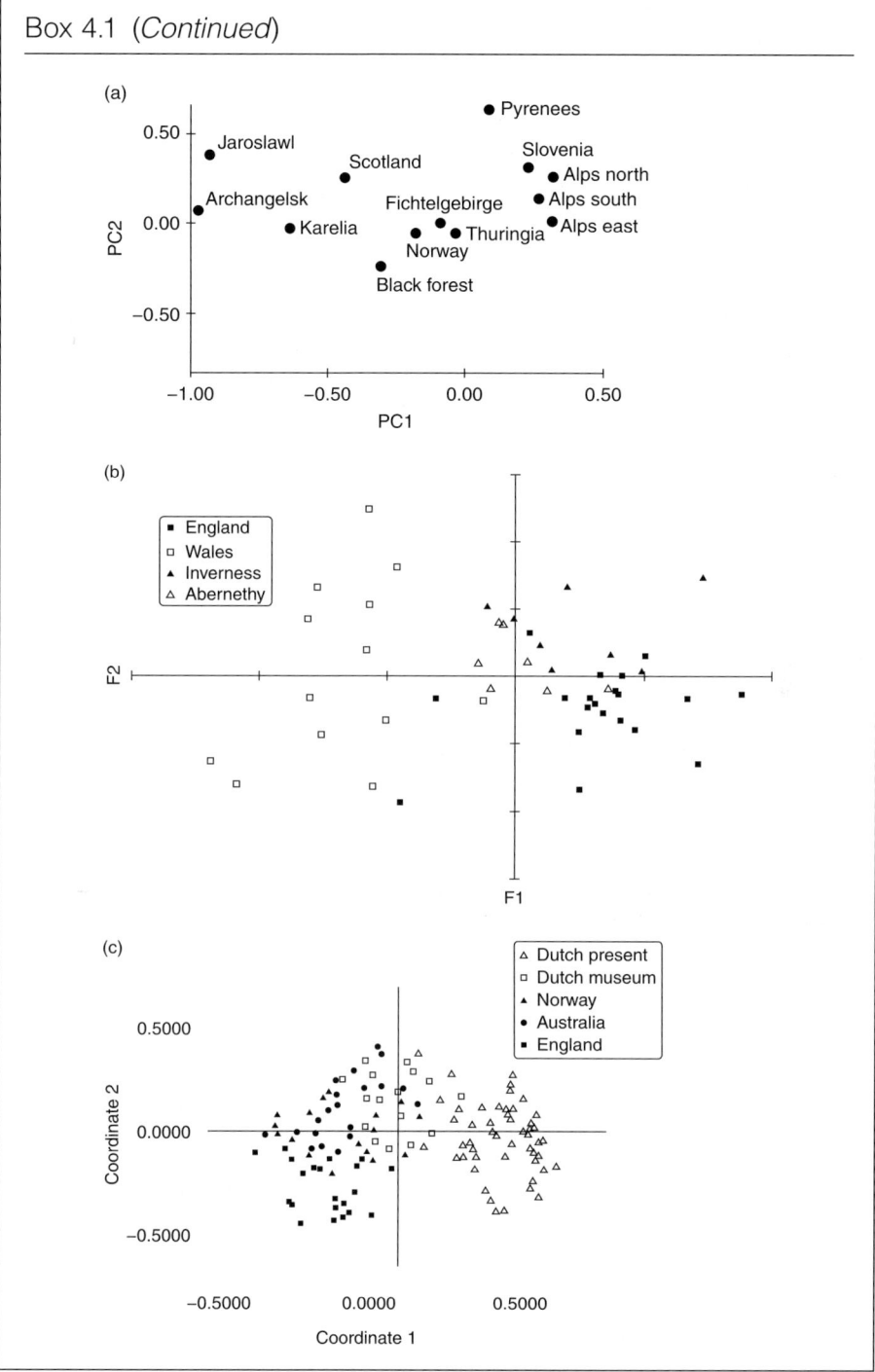

Box 4.1 (*Continued*)

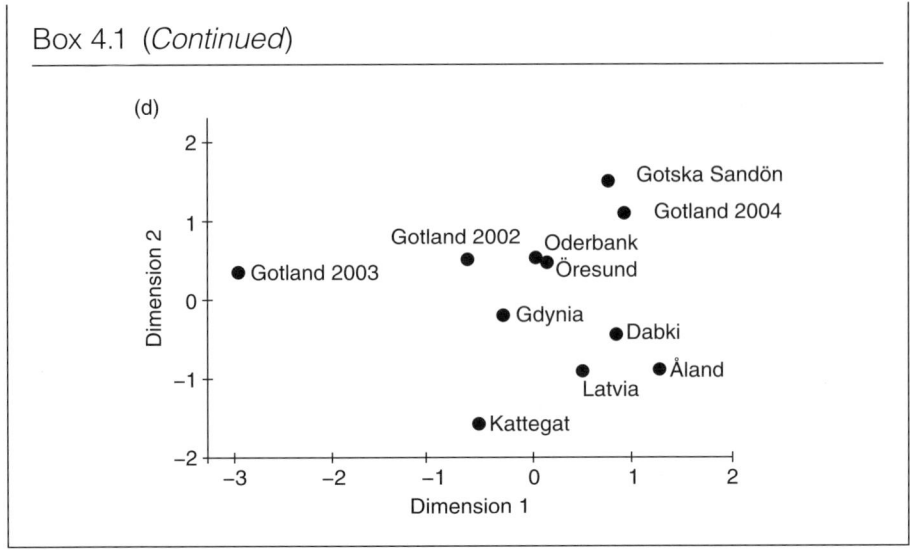

(PC1) in that populations in geographic close proximity tend to cluster together in the plot. Populations from Russia and Finland are found to the left in the figure (Box Fig. 4.1a) while populations from the Alps and the Pyrenees are found to the right with populations from central Europe in between. However, there is considerable scatter in the data and no evidence of clear breaks in the genetic structure in the form of separate clusters of populations (Segelbacher *et al.* 2003).

Another example may be provided by populations of black grouse, *Tetrao tetrix*. This species was once widespread in Great Britain but has faced contraction of its range, and remaining populations have become fragmented and increasingly isolated (Hudson and Baines 1993). At present, three separate distribution regions can be recognized. The first region is Scotland north of the Edinburgh–Glasgow belt. Here subpopulations are naturally fragmented since high mountains and lochs break up the suitable habitat for black grouse. The second region is around and in the northern Pennines, in England, and southern Scotland. The third region is in northern Wales. Populations in England/southern Scotland and Wales are smaller and more fragmented than those in northern Scotland, due to intense land management and areas of high anthropogenic activity. It is unlikely that there is much present connectivity between the described regions and migration between them is therefore probably very restricted if it exists at all. A correspondance analysis (CA) plot of genetic variation showed three clusters with some overlap when plotting the first two axes, accounting for 8.87 and 7.82% of the genetic variation, respectively (Box Fig. 4.1b). This plot suggests that discontinuities in the genetic data correspond with discontinuities in the geographic distribution. This conclusion is backed up by pairwise F_{ST} comparisons between the three regionally defined clusters which revealed significant genetic distances between all the regions included in the study (England compared with Wales $F_{ST} = 0.11$, England compared with Scotland $F_{ST} = 0.07$, and Wales compared with Scotland $F_{ST} = 0.10$; in all comparisons $P < 0.01$), whereas the pairwise distance between Abernethy and Inverness in Scotland was non-significant ($F_{ST} = 0.05$) after Bonferroni correction (J.K. Larsson *et al.*, unpublished results).

Presumably these differences have been exaggerated by the recent isolation of the populations among the three regions. However, it cannot be ruled out that the differences among regions simply depict isolation by distance and thus past clinal variation in gene frequencies. In another analysis using central European black grouse, we were able to show that geographic genetic differences among populations may evolve quickly. Using museum samples and samples from the only remaining black grouse population in the Netherlands, we demonstrated that the present populations have evolved to become different from other European black grouse populations in the course of the last 50 years (Box Fig. 4.1c).

However, temporal differences may also complicate geographical analyses. This is well known in studies of fish populations where cohort effects have been well studied. In turbot, *Psetta maxima*, three temporal samples from the same geographic location (off Gotland) showed as much variation as the whole geographic sample (Box Fig. 4.1d). Similarly, a previous report on geographic structure among European eel populations (*Anguilla anguilla*; Wirth and Bernatchez 2001) may be explained by temporal differences among age classes rather than geographic structure (Dannewitz *et al.* 2005).

It is often useful to be able to identify migrants among subdivided populations to estimate gene flow and connectivity. To this end several statistical methods that belong to the class known as assignment tests have been developed (Paetkau *et al.* 1995, Rannala and Mountain 1997, Cornuet *et al.* 1999). These methods use the *a priori* knowledge of source populations for the assignment of individuals to populations and are used to infer migration levels, isolation, and conservation status of several threatened species.

Paetkau and coworkers (1998) used multidimensional scaling and an assignment test to show that the large-bodied brown bears, *Ursus arctos*, of coastal Alaska were part of a continuous continental distribution of brown bears, and not genetically isolated from the physically smaller 'grizzly bears' of the interior of Alaska. By contrast, they found that the bears at Kodiak Island to the south of Alaska showed evidence of little or no genetic exchange with continental populations in recent generations (Fig. 4.1). It appears as though water is a dispersal barrier in brown bears, since data from the four insular populations indicated that dispersal can be reduced or eliminated by water barriers of as little as 2 km in width and that all individuals from Kodiak Island were assigned to the island. In the whole sample, bears could be correctly assigned to their population in 92% of the cases and there was a strong tendency for misassigned individuals to be assigned to the closest neighbouring study areas. This indicates that dispersal in brown bears occurs in a stepping-stone fashion.

In North America, wolverines, *Gulo gulo*, once occupied a continuous range from Alaska southward to New Mexico. In USA excluding Alaska, small remnant populations remain only in the northwest where they are connected to healthy populations in Canada. Assignment tests revealed a high degree of population substructure and low levels of gene flow in Montana (Cegelski *et al.* 2003). These results contrast to those from studies in the less fragmented landscapes of Alaska and Canada and suggest that wolverine populations of Montana are becoming increasingly fragmented due to human development and disturbance.

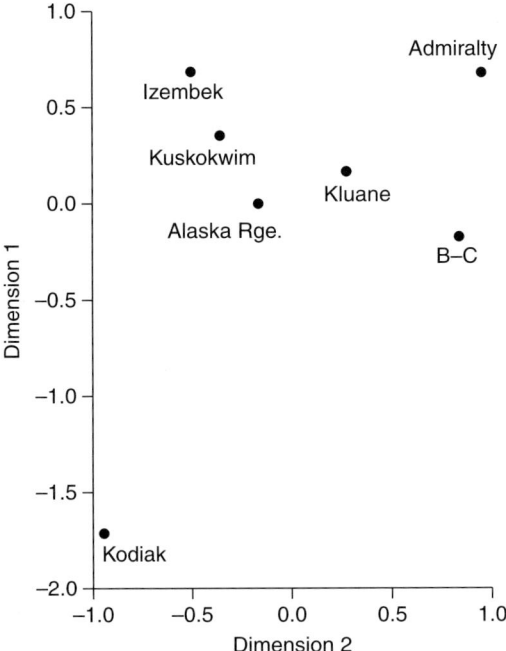

Figure 4.1 Cluster analysis of genetic distances between study areas using multidimensional scaling in Alaskan brown bears (from Paetkau *et al.* 1998, reprinted with permission from the publisher).

This wolverine study serves as a nice example of recent developments in the population genetic tools used in studies of human-induced population fragmentation. The assignment tests used in the study was the one originally developed by Paetkau *et al*. (1995). This test uses the observed allele frequencies for each of the predefined reference populations to calculate the likelihood of every observed genotype in each of the populations. Individuals are assigned to the population with the highest likelihood score (Paetkau *et al*. 1995).

Predefining populations was not straightforward in the case of the wolverines (Fig. 4.2). Therefore, the algorithms implemented in the software Structure (Box 4.2) were used to find the most likely number of populations given the data. Then, several different assignment methods including the one made possible by Structure were run. A high degree (84%) of concordance in population grouping was observed between methods, and individual assignments of the iterative method agreed with the results of Structure in 97% of the samples. The results suggest that wolverines in Montana are fragmented because of human infrastructure and disturbance and more so than further north in the species' range.

Studies of pumas, *Puma concolor*, in Utah, Colorado, Arizona, and New Mexico have identified a north–south divide in genetic structure (McRae *et al*. 2005). This division may be explained by a combination of old natural and new human-induced dispersal barriers. Although narrow habitat corridors appear to connect the northern and southern regions, these corridors are bisected by natural barriers to gene flow such as inhabitable grasslands and deserts, the Colorado River, and the Grand Canyon. However, and metropolitan areas and major roads may also impede movement between the north and south.

Likewise, research on bighorn sheep, *Ovis canadensis* (Epps *et al*. 2005), bobcats, *Lynx rufus*, and coyotes, *Canis latrans* (Riley *et al*. 2006), lowland populations of the common shrew, *Sorex araneus* (Lugon-Moulin and Hausser 2002), roe deer, *Capreolus capreolus* (Wang and Schreiber 2001, Coulon *et al*. 2006), and winter moth, *Operophtera brumata* (Van Dongen *et al*. 1997), show that human-induced habitat fragmentation, roads, and other anthropogenic barriers may block gene flow and cause rapid declines in genetic diversity.

In fish, the erection of hydroelectric power dams and other changes to rivers and lakes have had dramatic effects of gene flow and population differentiation. In salmonids, decreased genetic diversity has been found when populations have become isolated (Carlsson and Nilsson 2001, Castric *et al*. 2001, Costello *et al*. 2003, Taylor *et al*. 2003, Wofford *et al*. 2005).

Human-induced changes may not only increase isolation and impair gene flow. Studies of three *Rorippa* species (Brassicaceae), *Rorippa amphibia*, *Rorippa palustris*, and *Rorippa sylvestris*, in northern Germany provide evidence for different patterns of gene flow in natural and in anthropogenic environments (Bleeker and Hurka 2001). It is argued that landscape changes in north-western

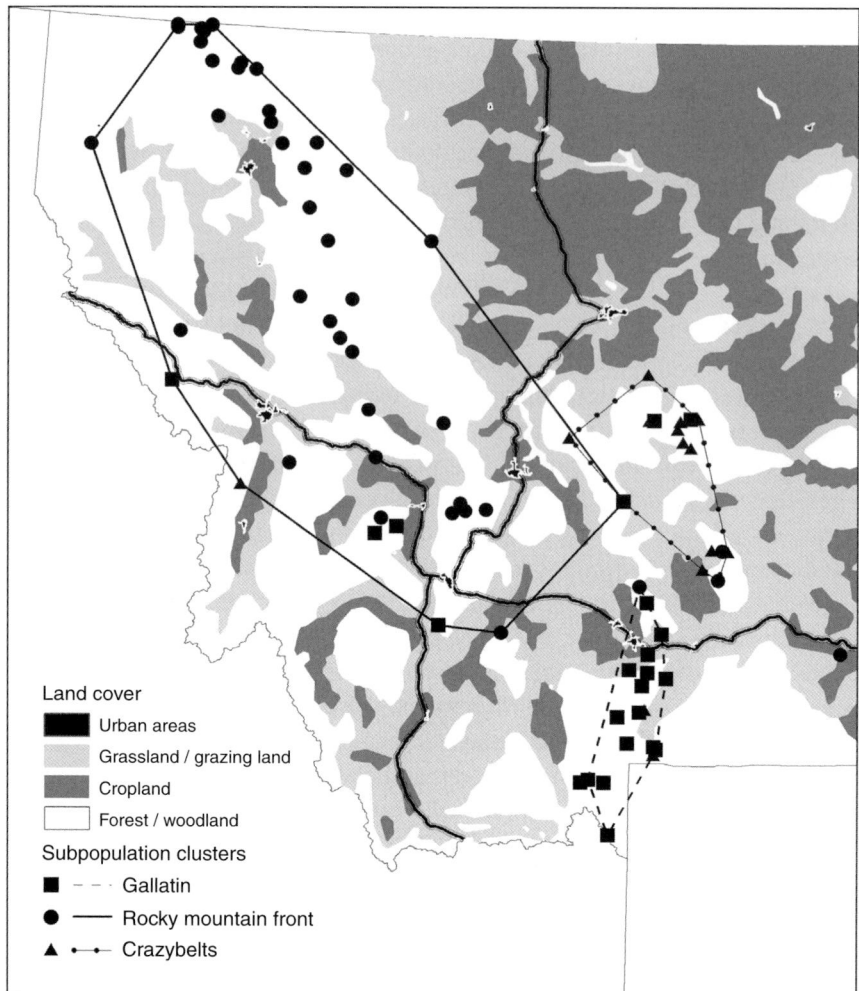

Figure 4.2 Map of Montana, USA, showing genetic groupings of individual wolverines as determined by Structure software along with land cover, major cities, and interstate highways (from Cegelski *et al.* 2003, reprinted with permission from the publisher).

Germany, in particular the creation of drainage ditches, has altered the patterns of gene flow and formation of hybrids between these species. This has had consequences for the ecotypic differentiation within *R. amphibia*.

The model-based algorithms in programs such as Structure, Partition, and Baps (Pritchard *et al.* 2000, Dawson and Belkhir 2001, Corander *et al.* 2003) thus allow studies of populations in which it is difficult to predefine populations. These programs infer the number of independent genetic clusters in a sample but

Box 4.2 **Bayesian inference of population structure**

Several software programs have implemented a model-based Bayesian approach to infer population structure without a definition a priori. The most common programs used are Structure (Pritchard *et al.* 2000), Partition (Dawson and Belkhir 2001), and Baps (Corander *et al.* 2003). Under varying assumptions about the number of putative number of populations (K), the likelihood of the data and posterior probabilities for different values of K may be calculated. Furthermore, the samples included in the study may be assigned with varying probability to any of the clusters detected.

The most widely used Bayesian approach is implemented in Structure in which the model applied accounts for the presence of Hardy–Weinberg or linkage disequilibrium by introducing population structure and attempts to find population groupings that (as far as possible) are not in disequilibrium (Pritchard *et al.* 2000). The estimated log probability of the data $\Pr(X|K)$ for each value of K among runs can then be compared (Pritchard *et al.* 2000). This allows for an estimation of the most likely number of clusters.

Evanno et al. (2005) tested the ability of the algorithm used in Structure to detect the true number of clusters (K) in a sample of individuals when patterns of dispersal among populations were not homogeneous. The authors used various dispersal scenarios from simulated data and found that the estimated log probability of the data $\Pr(X|K)$ did not provide a correct estimation of K. However, when they used a new statistic that they developed and called ΔK, which was based on the rate of change in the log probability of data between successive K values, they were able to better retrieve the true value of K.

have also been used extensively as assignment tests, in population admixture and hybridization analysis, migration and dispersal analysis, and also in attempts to detect cryptic genetic structure of natural populations (see references in Höglund and Shorey 2003, Evanno *et al.* 2005).

4.2 Landscape genetics

Landscape genetics has emerged from a combination of spatial statistics, molecular genetic techniques, and landscape ecological theories (Manel *et al.* 2003, Holderegger and Wagner 2006). This approach uses individuals as the study unit and attempts to address whether geographic and environmental structures affect gene flow and genetic structure. In an early study using this approach, it was suggested that rivers and unsuitable habitat in Scotland can prevent gene flow in red grouse, *Lagopus lagopus scoticus* (Piertney et al. 1998). Similarly, large rivers and mountains seem to have prevented gene flow and contributed to species and subspecies formation in the superspecies *Manacus* (Höglund and Shorey 2004). Some landscape genetic approaches use algorithms very similar to the model-based approaches reviewed above to find population structure, and thus no *a priori* assumptions about population structure are required. The difference, for example, to the algorithms in the program Structure is that landscape genetics directly allows inference of population structure based on geo-referenced genetic data (Guillot et al. 2005, Holderegger and Wagner 2006).

In landscape genetics, spatial statistical analyses of population structure can be coupled with geographical information system (GIS) analyses to provide insight to the influences of the environment on evolutionary genetic processes. GIS data may include topography (i.e. slope, elevation, and distance), habitat type, ground moisture levels, and bodies of water. Correlations of these landscape variables to genetic differentiation can be quantified and the information applied to identify likely dispersal routes and barriers to gene flow.

Guillot *et al.* 2005 introduced a spatial statistical model which provides the power to infer and locate genetic discontinuities between populations using individual geo-referenced multilocus genetic data. Their Bayesian model includes Markov chain Monte Carlo simulations to infer the spatial model parameters. The model can locate genetic discontinuities including cryptic spatial genetic structure, estimate the number of populations for a sample area, quantify the spatial dependence in the data set, detect migrants, and assign individuals to their population of origin. One shortcoming of the spatial statistical model, however, is the inability to separate spatial dependency, due to processes such as kin clustering, isolation by distance, and selfing, from true genetic discontinuities. Population numbers could be overestimated in study species that include these processes, especially if the life-history traits of the study organism are not documented or are unclear.

Studies of a range of organisms have used a combination of the landscape genetic approach and genetic estimates of dispersal distance to infer levels of population structure and gene flow in wild animals (e.g. roe deer, *Capreolus capreolus*, Coulon *et al.* 2006; otter, *Lutra lutra*, Janssens *et al.* 2008). Five major research categories to which landscape genetics can be applied have been identified: (1) quantification of how observed genetic variation is influenced by landscape variables and layout, (2) identification of gene-flow barriers, (3) identification of source–sink dynamics and migration routes, (4) understanding the spatial and temporal scale of ecological processes, and (5) species-specific hypothesis testing (Storfer *et al.* 2007).

Spear *et al.* (2005) investigated genetic diversity and structuring of the blotched tiger salamander (*Ambystoma tigrinum melanostictum*) across 10 sites of its northern range using eight microsatellite loci. The authors examined how various landscape variables are correlated with genetic differentiation in this sample area and tested multiple hypothetical dispersal routes against a null model. Gene flow was found to be highly restricted among sites. A straight-line topographic model best estimated dispersal routes with river crossings and open shrub habitat seemingly supporting increased gene flow whereas distance and elevation apparently increased differentiation. These results were somewhat surprising and contrary to predictions (Fig. 4.3).

The authors predicted that a stepping-stone, least-cost habitat or least-slope dispersal model would perform best as amphibians are expected to travel through preferred wetland habitat and avoid increased slope and elevation change (Funk *et al.* 2005). However, neither stepping stone, wetland likelihood, nor least-slopes models explained more variation in the genetic data than the straight-line model. Furthermore, although rivers were predicted to obstruct gene flow due to the presence of predatory fish, they were in fact positively correlated with decreased population differentiation. The authors suggest that the observed relationship of rivers and gene flow may be due to drought in the sample area: the river routes may be preferable to the desiccated surrounding habitat devoid of standing water. Another counterintuitive finding was that open habitat facilitated whereas closed forest cover decreased gene flow. This is contradictory to expectations as open areas are thought to limit dispersal in salamanders (Madison and Farrand 1998, de Maynadier and Hunter 1999, Rothermel and Semlitsch 2002). The open habitat, in this case, mostly included previously burned areas with some vegetation regrowth: apparently burned areas facilitate movement. This GIS and spatial statistical analysis of genetic data provided new information for the conservation and management of the blotched tiger salamander. Some previously determined barriers to gene flow were discounted and several counterintuitive facilitators of gene flow were identified. The study also introduced a possible new technique to support increased mobility between these sample populations: prescribed burning.

In a study of hazel grouse, *Bonasa bonasia*, my research group used 613 geo-referenced tissue samples from northern Sweden where each individual was genotyped at 12 microsatellite loci, to make inference on population genetic structure, gene flow, and dispersal (Sahlsten *et al.*, 2008). Using a spatial statistical model for landscape genetics to infer the number of populations and the spatial location of genetic discontinuities between putative populations, we found indications that Swedish hazel grouse are divided into northern and southern populations. We could not find a sharp border between these two populations and none of the observed borders appeared to coincide with any potential

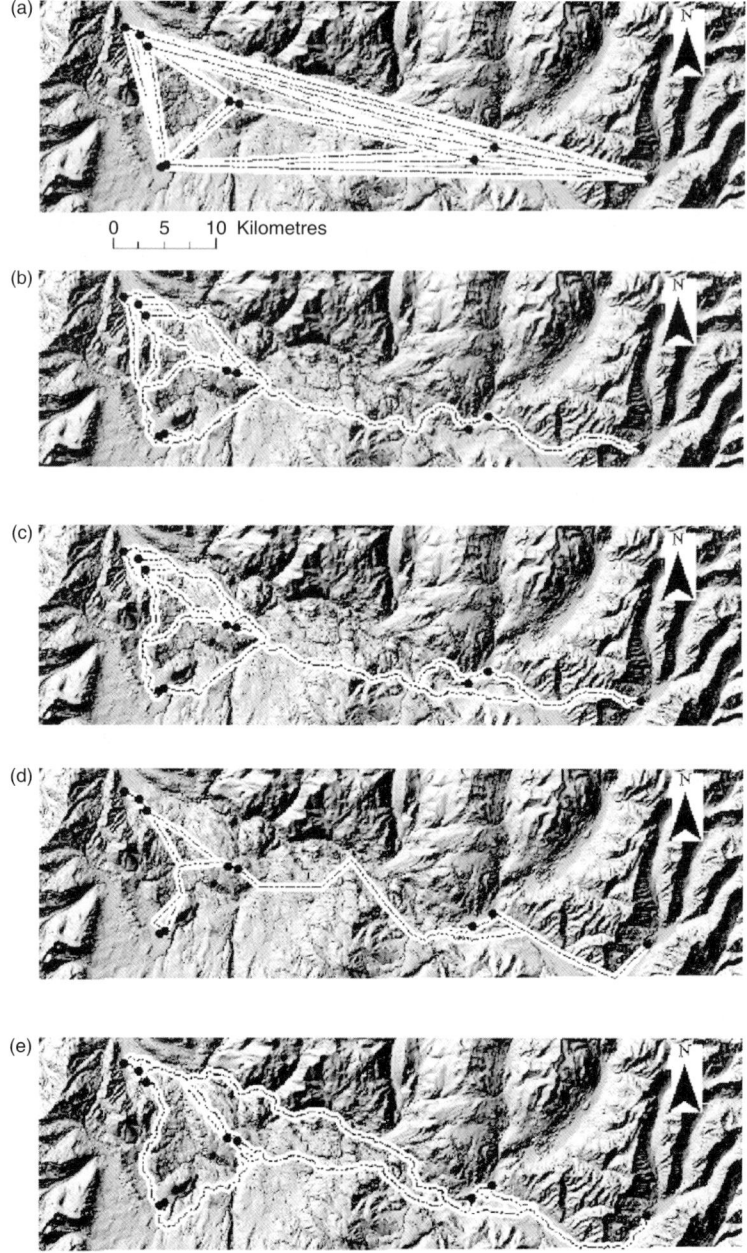

Figure 4.3 Maps representing model routes for salamander gene flow across a landscape in the northern USA. The background is a shaded relief map of the study area. (a) Straight-line route; (b) wetland likelihood route; (c) combination of least-slope/wetland likelihood; (d) stepping-stone route; (e) least-slope route (from Spear *et al*. 2005, reprinted with permission from the publisher).

geographical barriers. These results imply that gene flow appears unrestricted in the boreal taiga forests of northern Sweden and that the two populations of hazel grouse in Sweden may be explained by the post-glacial reinvasion history of the Scandinavian peninsula rather than any present-day physical hindrances to gene flow. Such hindrances were present, however, when the Scandinavian peninsula was recolonized after the last glaciation. The block of inland ice that melted away last of all (excluding present-day glaciers) was situated in the centre of the Scandinavian peninsula. This had the consequence that since Scandinavia was recolonized from two directions: the south west and the north east, respectively, many organisms, including modern humans, now inhabiting Scandinavia show evidence of two genetic populations with a contact zone positioned approximately in central Sweden (e.g. bears, *Ursus arctos*, Taberlet *et al.* 1995; willow warblers, *Phylloscopus trochilus*, Bensch *et al.* 2002; shrews, *Sorex araneus*, Andersson 2004). Thus we could detect no presence of present-day natural or human-induced barriers to gene flow but we could detect the ghost of a barrier in the past.

Segelbacher *et al.* (2008) used landscape genetics to analyse individual genetic variation in capercaillie *T. urogallus* in the Black Forest mountain range in south-western Germany. Due to human-induced habitat fragmentation, Black Forest capercaillie has declined rapidly during the last decades and now persists in patchy isolated fragments. Despite overall low genetic structure, the authors found strong indications for a major boundary separating the northern part of the Black Forest area from the other subpopulations. Among historic samples, genetic differentiation was very low, indicating that the current genetic structure is caused by recent habitat fragmentation.

4.3 Effects of bottlenecks and how to detect them

What happens to genetic diversity when populations get contracted in numbers? Previous chapters have reviewed the evidence and showed that in general terms genetic diversity is lost. However, the details of this loss of diversity affect the different measures of genetic variation in different ways. The most well-known difference is that during a population bottleneck alleles are lost faster than heterozygosity (Watterson 1984, Maruyama and Fuerst 1985). This difference has the consequence that the impacts on genetic diversity of a population size reduction can be estimated from data in extant populations by examining patterns of heterozygosity excess and observed allele frequencies without knowledge of the genetic variation in the past (Cornuet and Luikart 1996, Piry *et al.* 1999). By examining the difference between the expected heterozygosity under Hardy–Weinberg equilibrium (H_e) and the heterozygosity expected at mutation–drift equilibrium (H_{eq}), inferences about losses of genetic variation can be made. In populations that have not been reduced in numbers and that are near mutation–drift equilibrium, H_{eq} will equal H_e (Luikart and Cornuet 1998). As alleles are lost more rapidly than heterozygosity during a population-size reduction the effect will be a heterozygosity excess (higher H_e) in reduced populations. Threatened species are rarely assayed continuously during population size reductions for genetic diversity. This method, which is implemented in the software Bottleneck (Piry et al. 1999), has appeal since it allows detection of loss of genetic variation by only requiring a single 'snap-shot' point estimate. However, theory predicts that a new mutation–drift equilibrium may be set rapidly when effective population size becomes low (Watterson 1984).

Populations of British natterjack toad, *Bufo calamita*, vary in the extent to which they have been reduced in numbers. Bottleneck tests were applied to microsatellite allele frequency data from these populations and the outcomes were compared with demographic information (Beebee and Rowe 2001). The tests correctly identified the populations in which bottlenecks have occurred and it was concluded that the approach was useful in demonstrating whether amphibian declines have occurred and could be applied to cases where long-term demographic time series are not available. Similarly, signs of bottlenecks using genetic data were detected in tiger salamanders, *A. tigrinum* (Spear et al. 2006), a population of Japanese macaques, *Macaca fuscata* (Kawamoto et al. 2007), silver rice rat, *Oryzomys argentatus* (Wang et al. 2005), Mediterranean monk seal, *Monachus monachus* (Pastor et al. 2004), Barbary red deer, *Cervus elaphus barbarus* (Hajji et al. 2007), and pine trees, *Pinus taeda* (Al-Rababah and Williams 2004).

Studies on black grouse, *T. tetrix*, on the other hand, suggest that this snap-shot approach may also fail to detect loss of genetic variation. Despite a decline in numbers from over 10 000 birds to fewer than 30 over the last 50 years in the last remaining population in the Netherlands, we could not detect any evidence of a bottleneck using the snap-shot approach (Larsson et al. 2008). One explanation for this may be that the effects on heterozygosity after a population crash will only persist for $(0.2–4)N_e$ generations before a new equilibrium is set (Maruyama and Fuerst 1985, Luikart and Cornuet 1998). In the studied population, there was a decline in numbers from about 7500 to about 1000 individuals between the 1950s to the 1970s. This would lead to a measurable heterozygosity excess for the population for about 67–200 generations, if the population would have remained constant at that size. However, with a census size of about 20 males the 5 years before the sampling of the extant population, a conservative estimate of effective population size is about 13 ($N_e = 4N_f \times N_m/(N_f + N_m)$, $N_f = 20$ and $N_m = 4$). This is under the reasonable assumptions that only a fraction of the males in this lekking species mate and that the sex ratio is 50:50. If so, a new equilibrium ($H_e = H_{eq}$)

has been set (0.2 × 13 = 2.6 generations) and the only way to suspect that such a population has been subjected to severe genetic drift is by comparison with other continuous populations or to access samples from prior to the population crash. Fortunately, such data were available in this case and by comparison with the genetic diversity observed in museum samples taken prior to the bottleneck, we could show that Dutch black grouse have indeed lost both alleles and heterozygosity (Larsson *et al.* 2008).

Similarly, in banner-tailed kangaroo rat, *Dipodomys spectabilis*, populations from south-eastern Arizona, USA, that were known to have experienced recent demographic reductions, genetic analyses with eight microsatellite loci failed to detect any bottleneck signals. The authors ascribed this failure to the populations being connected by dispersal and suggested that bottlenecks may be difficult to detect using molecular genetic data in systems with extensive dispersal (Busch *et al.* 2007).

Island colonization and founder effects were studied in introduced ship rat populations of *Rattus rattus* in the Guadeloupe Archipelago (Abdelkrim *et al.* 2005). Three different methods to detect bottlenecks were tested. These where the heterozygosity excess, the mode-shift indicator (Piry *et al.* 1999), and the *M* ratio (Garza and Williamson 2001) methods. The heterozygosity excess and the mode-shift indicator only detected bottlenecks for the recent colonization on two of the islands. However, bottlenecks were detected for all the populations using the *M* ratio method. Taken together, all studies that fail to detect bottlenecks despite good evidence that a bottleneck has indeed taken place suggest caution when applying these tests. The assumptions behind the tests need to be fulfilled. At small population size a new mutation–drift equilibrium is rapidly set and the difference between H_e and H_{eq} disappears.

4.4 Effects of population expansions and range shifts

As in the case of contracting populations, the genetic patterns expected if a population suddenly increases in number from a very small size are different from what would be expected at genetic equilibrium. The number of alleles in an expanding population is elevated over what is expected in a population at mutation–drift equilibrium showing the same expected heterozygosity (Maruyama and Fuerst 1984). This is what would happen in a newly colonized area where positive population growth is possible or when a population has recovered from a severe population size bottleneck.

What happens when species are forced to move? This is a question of growing concern as climate change has been identified as one of the major threats to global biodiversity in the near future (Parmesan and Yohe 2003, Root *et al.*

2003). So-called climate envelope models have been developed to predict the changes and possible future extinctions of present-day biota (e.g. Townsend-Peterson *et al.* 2002, Thomas *et al.* 2004). Such models are using present-day distributions to calculate a climate envelope of the species distribution and then use predicted changes in climatic variables to project the future distribution under varying assumptions. Depending on the size and extent of the future climate envelope, risk assessments about global and local extinction can be made. These models often predict loss of biodiversity even under conservative scenarios.

One implicit assumption of envelope models is that the envelope or 'niche' of the species studied is unchanged as the species is forced to move or change distribution. That is, there is no microevolutionary change allowed in these models. As the climate becomes warmer, poleward range shifts of species, communities, and ecosystems are predicted worldwide. The response of species to changing environments is likely to be determined largely by microevolutionary responses in populations at the range margins (Hampe and Petit 2005).

Hampe and Petit (2005) summarized the processes likely to be important when the range of a species is shifted towards one of the poles when the climate becomes warmer (Fig. 4.4). In the expanding edge, chance is likely to be involved during founder events and whether or not the conditions in the new range favour positive population growth. Furthermore, if a species becomes established, traits like dispersal ability and cold stress tolerance are likely to be favoured. At the

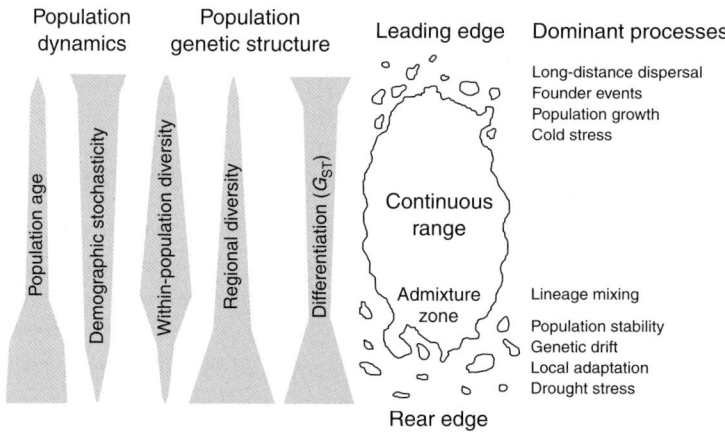

Figure 4.4 Population features and population processes at the leading and rear edges of species ranges. The width of grey bars shown on the left indicates the relative importance at the corresponding position within the range. G_{ST} is an F_{ST} analogue (from Hampe and Petit 2005, reprinted with permission from the publisher).

low-latitude limit (rear edge) the importance of genetic drift will increase as the populations become smaller and more fragmented and the possibilities for local adaptation to selective pressures such as drought stress become harder. In the rear of the continuous range there will be an area of population stability with an admixture zone where different lineages evolving in the rear-end refuges may mix. Thus this scenario predicts different outcomes on genetic variation and the processes involved shaping that variation. Thus genetic variation will be low both within and among populations at the leading edge as only genotypes preadapted to dispersal will be favoured. At the centre of the continuous range diversity at all levels will be moderate. At the rear end, the within population diversity will be low but regional diversity and population differentiation (as measured by F_{ST}) will be high.

The genetic consequences of range expansion in an expanding moss species, *Pogonatum dentatum*, were studied by comparing source populations in a mountain area with populations from a recently colonized lowland area in Sweden (Hassel *et al*. 2005). As expected, genetic variation was lower in the newly colonized populations in which three out of four populations showed evidence of having passed a bottleneck as would be predicted by the populations being formed after recent founder events. However, the newly founded populations showed higher haplotype diversity, less linkage disequilibrium, and fewer compatible loci. This indicates that sexual recombination is more important in the newly colonized populations than in the source population. The authors suggest that a higher success of establishment from spores occur in the new areas whereas clonal propagation predominates in the source populations. As predicted by Hampe and Petit (2005) there was less genetic differentiation among the newly colonized populations than among source populations. However, the authors attributed this to more extensive gene flow involving more spores moving among populations in the leading-edge populations.

What makes a good colonist? Among ecological traits the following may be identified: good dispersal abilities, high population growth potential, possibilities for both clonal and sexual reproduction, being adapted to ephemeral and unstable habitats, and, in the case of animals, being an generalist predator (see pp. 210–212 in Newton 2003). When it comes to the issue of genetic variation it may be argued that small invading populations with low genetic variability face the same problems as contracting threatened populations. However, as argued above, invading (leading-edge) and contracting (rear-edge) populations may differ genetically in many respects. In expanding populations the larger and more genetically diverse any colonizing propagule is, the larger the chance of a successful invasion (Lenormand 2002). Repeated invasions to the same location from different sources may also boost genetic variation in the new habitat and affect the probability that a colonizing species may become firmly established

(Kolbe *et al.* 2004). Among these traits, patterns and rates of migration, effective population size, and number of pioneer individuals (i.e. founder events) may be estimated using molecular markers (Estoup *et al.* 2004).

As gene flow tends to work against local adaptation (see Chapter 6), this may limit the geographic range of any expansion (Kirkpatrick and Barton 1997). On the other hand, gene flow increases the genetic variance of local populations, which is the necessary raw material for natural selection to produce locally adapted genotypes. When population size becomes small, the importance of drift increases and under such circumstances dominance and epistatic variance may be converted into additive genetic variance (Cheverud and Routman 1996, Reznick and Ghalambor 2001).

4.5 Invasive species

From expanding species the step is not far to invasive species. Invasive species have been defined by the Australian government as 'a species occurring, as a result of human activities, beyond its accepted normal distribution and which threatens valued environmental, agricultural or other social resources by the damage it causes' (www.environment.gov.au/biodiversity/invasive/). This definition stresses that humans are the cause of the invasion. Other definitions lack this criterion, for example: 'introduced species cause negative impacts on the environment, human activities, or human health' (Lee 2002). Invasive species have been identified as major threats to biodiversity when they become aggressive and exact a toll on ecosystem diversity and processes (see papers in Mooney and Hobbs 2000, Lee 2002). Not all introduced species become pests and have negative impacts on the local flora and fauna. However, some do become serious threats to native biodiversity. What makes some species become 'aggressive' and others not? In plants it has been suggested that in native areas plants have coevolved with enemies such as herbivores and competitors. A plant that has become a pest may just have been lucky and released from these enemies. As a consequence it can grow unchecked. This enemy-release hypothesis may thus explain the rapid increase in distribution and abundance of some plants.

Evolutionary aspects of species invasiveness have been neglected in past research (Mooney and Cleland 2001, Lee 2002). However, recent studies provide evidence that the success of many invaders depend to a larger extent on their ability to respond to natural selection than on their physiological tolerance or phenotypic plasticity (Lee 2002). The papers reviewed by Lee (2002) highlighted two findings: epistatic interactions among genes may contribute to adaptation during invasions and the effects of a small numbers of genes could have profound effects on invasion success.

The perhaps most famous example of an invasive species and without doubt one of the best well-known cases from a genetic point of view is that of the cane toad, *Bufo marinus*. This species was introduced to Australia from its native range in north-east South America (via a range of islands) in 1935 (Easteal 1981, Lever 2001). In Australia, it has proliferated and is still spreading (Fig. 4.5). The expansion history since 1960 in two areas of Australia have been studied with the aid of microsatellite and allozyme markers (Estoup *et al.* 2001, 2004). Bayesian estimation of various demographic models applied to the genetic data suggests that the effective number of migrants appears to be considerably lower than that of founders in both expansion areas (Estoup *et al.* 2004).

The cane toad has had adverse effects on Australian biodiversity (Phillips *et al.* 2003). There is also evidence of evolutionary change in both the invasive cane toad populations and the biota with which they interact. Cane toad morphology has changed during the invasion, where leg length have become relatively greater, presumably as a response to selection on migratory speed (Phillips *et al.* 2006a). Furthermore, two gape-size-limited snake species have evolved smaller gapes to avoid the toxic toads (Phillips and Shine 2004) and

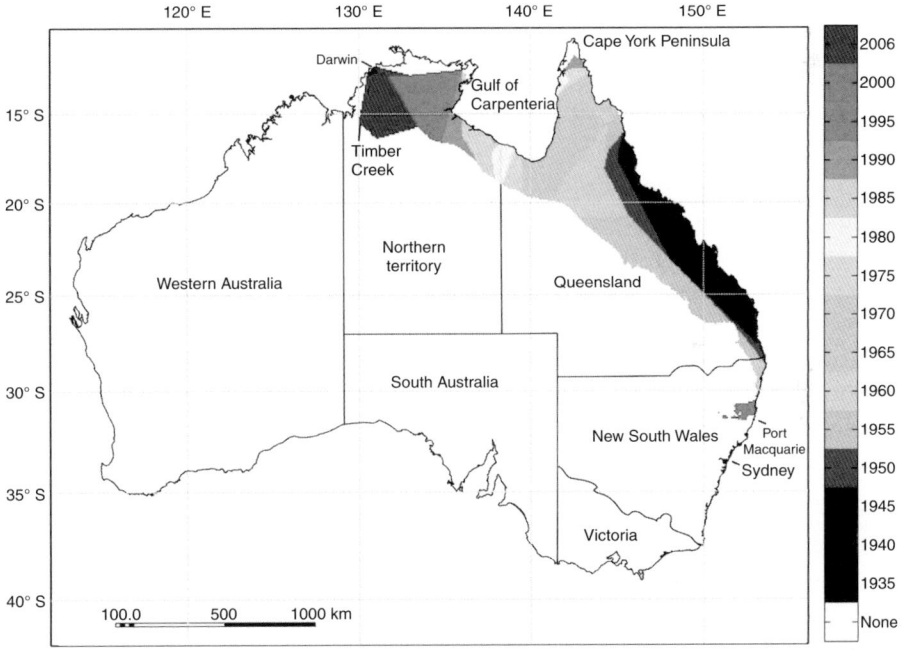

Figure 4.5 Map of Australia showing the change in the range of cane toad range in 5-year increments (6 years for the latest estimate). Key cities and geographic features are indicated (from Urban *et al.* 2008, reprinted with permission from the publisher).

Australian black snakes, *Pseudechis porphyriacus*, from toad-exposed localities showed increased resistance to toad toxin and a decreased preference for toads as prey (Phillips *et al.* 2006b).

The Argentine ant *Linepithema humile* is native to South America and has spread widely across the globe. In both the USA and New Zealand the introduction appears to coincide with a change in the social system of the ants. In introduced areas there is a widespread genetic similarity among colonies as a result of the colonies being formed by a few founding individuals. Relatedness within nests and colonies is lower in the introduced range in the USA than in the native range in Argentina, where intra-colonial relatedness is high and colonies are genetically differentiated (Tsutsui and Case 2001). The New Zealand population of Argentine ants is also characterized by low levels of genetic variation and no signs of population differentiation or isolation by distance among colonies could be found (Corin *et al.* 2007). These differences between the native and introduced areas appear to coincide with behavioural differences. In introduced areas the levels of aggression among ants is low which is thought to facilitate the invasiveness and spread of the ants in their new areas. These findings show that the introduction of a species to a new area can have dramatic consequences not anticipated by studies in the native area.

4.6 Summary

This chapter has reviewed the genetic consequences of changes in the environment. These changes are often so rapid that contemporary populations are often not found in genetic equilibrium. Furthermore human-induced habitat fragmentation often results in a complex mosaic of remaining populations that differ in size and connectivity. Fortunately, a number of tools have been developed to detect population structure, gene flow, and evolution in such complex situations. Furthermore, this chapter has provided evidence of rapid evolutionary responses in many organisms to changes in the environment. Such changes may be induced by a multitude of factors, such as habitat loss and fragmentation, hindrances to dispersal and hence gene flow, climatic changes, and introduction of invasive species.

5 Genes under selection: Mhc and others

For good reasons workers in the field of molecular population genetics have by tradition used neutral genetic markers to study evolutionary processes. By being able to ignore selection, such markers have allowed estimation of the strength and importance of mutation and recombination, genetic drift, and migration in shaping genetic diversity among and within populations. However, variation at neutral loci cannot provide direct information on selective processes involved in the interaction between individuals and their environment, nor on the capacity for future adaptive changes (Meyers and Bull 2002, van Tienderen *et al.* 2002, Sommer 2005). The mechanisms that maintain and promote adaptive genetic diversity in natural populations is a central issue in evolutionary ecology and conservation (Orr and Coyne 1992, Hedrick 2001, Boake 2002, Sommer 2005). How adaptive genetic diversity is apportioned across both space and time provides insight into how adaptation may progress under novel or changing environmental conditions, and the extent to which populations may be prone to stochastic extinction through the erosion of genetic diversity. Such an endeavour would need the study of coding genes and the regulatory mechanisms that underlie adaptive phenotypes in natural populations.

That genetic diversity has been estimated by using neutral genetic markers was also partly driven by the fact that coding loci were hard to access in non-model species. In the past, it was often assumed that neutral and adaptive variation are correlated (Hedrick 2002). Although the relationship may sometimes hold, the correlation between neutral and adaptive genetic diversity is usually rather weak (Hedrick 2001) and sometimes even absent (e.g. Madsen *et al.* 2000).

This chapter will focus on genes under selection. Much of what is known about 'ecologically relevant' genetic variation at the level of DNA sequences comes from studies of genes of the major histocompatability complex (*Mhc* genes). This gene family codes for cell-surface proteins involved in immunoresistance in vertebrates. I will briefly, since there are a number of excellent reviews on this topic, review evidence for selection on *Mhc* loci, links to parasite resistance, and consequences of lost genetic variation at *Mhc* loci. Not all immune genes belong to the *Mhc* family and there is a growing concern that immunoecological studies should address other immunity genes (Acevedo-Whitehouse and Cunningham 2006). At the end of the chapter other candidate genes will also be covered. Examples of such that may be relevant in conservation are genes coding for animal pigmentation (such as *mc1r*) and *clock* genes (involved in photoperiodism).

5.1 *Mhc* genes

In 2006 Piertney and Oliver stated that 'our understanding of how selection can act to maintain adaptive polymorphism in natural populations remains based on a small number of key gene regions, such as the major histocompatibility complex (*Mhc*)'. This cluster of genes has been extensively studied in both model and non-model species during the last decades (see reviews by Brown and Eklund 1994, Apanius *et al.* 1997, Edwards and Hedrick 1998, Jordan and Bruford 1998, Penn and Potts 1998, 1999, Tregenza and Wedell 2000, Zelano and Edwards 2002, Bernatchez and Landry 2003, Garrigan and Hedrick 2003, Mays and Hill 2004, Ziegler *et al.* 2005, Piertney and Oliver 2006, Sommer 2005).

Mhc genes are among the best candidates for the study of adaptive genetic diversity as they are extraordinarily variable and of obvious ecological relevance. The cell-surface proteins encoded by *Mhc* class I are found on all cells and bind to epitopes from antigens derived from intracellular pathogens, such as viruses, and present these on the cell surface (Fig. 5.1). Class II molecules are only found on specialized immune cells, for example macrophages, that engulf extracellular parasites and bind epitopes derived from such extracellular pathogens. These can then be recognized by the helper cells that trigger the production of specific antibodies by B cells. In this process MHC class II molecules are involved in the signalling between B and T cells. Both molecules are therefore important in the triggering of the adaptive immune response. Hence there is a direct link between *Mhc* genes and individual fitness. Furthermore, vertebrate *Mhc* genes are among the most variable loci known in humans, with over 500 alleles found at a single locus (Robinson *et al.* 2003).

There is considerable variation in the organization and size of the *Mhc* among vertebrates. In humans, the *Mhc* complex contains 421 loci (Horton *et al.* 2004). In domestic chicken the classical region (BF/BL) is much smaller—about 20 genes—and is therefore sometimes referred to as the minimal essential *Mhc* (Kaufman *et al.* 1999, Kaufman 2000). Because the ability of MHC to bind to broad arrays of pathogens is related to a high allelic sequence variation in the region coding for the antigen-binding sites (Doherty and Zinkernagel 1975), this high level of polymorphism is likely to be maintained by balancing selection resulting from heterozygote or rare-allele advantage (Takahata and Nei 1990). In

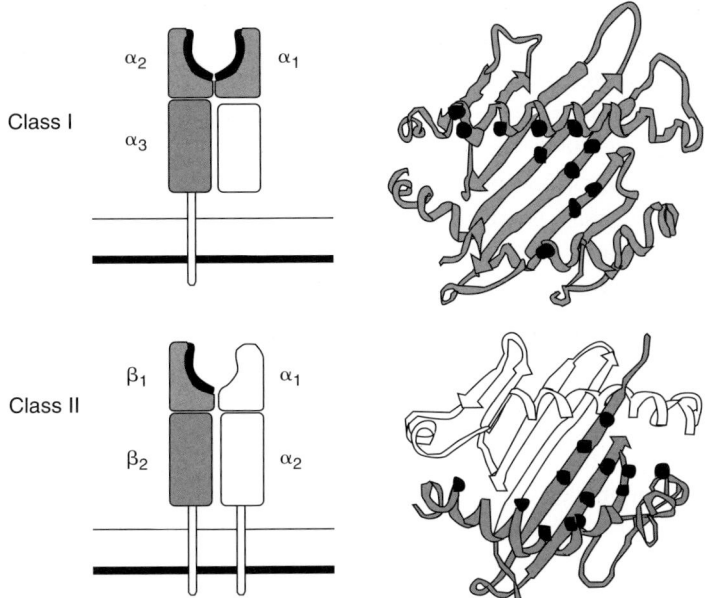

Figure 5.1 Schematic picture of MHC class I and II molecules. To the left, the molecules are seen from the side with the cell surfaces at the bottom. Antigen-binding sites are shown by the black areas and the approximate positions of α and β chains are indicated. To the right the molecules are shown from the top with the antigen-binding sites in black.

addition, MHC disassortative mating preferences (Landry and Bernatchez 2001, Penn 2002, Zelano and Edwards 2002, Milinski 2006), as well as prenatal foetal incompatibilities in mammals (Ober 1999), can contribute to the maintenance of extreme levels of polymorphism.

One approach to studying selection at *Mhc* loci has been to identify balancing selection in the current generation. The tools used have been observed deviations from Hardy–Weinberg equilibria, Mendelian expectations, or expectations about random associations (Garrigan and Hedrick 2003). Furthermore, associations have been looked for between specific genotypes and fitness on exposure to certain environments. Associations between specific *Mhc* alleles and disease resistance or susceptibility have been found in a number of species including humans (Sommer 2005). When looking at *Mhc* evolution over evolutionary times the most common approach has been to examine the ratio of non-synonymous to synonymous substitutions (dN/dS) in the sequences coding for the molecule. There are two hypotheses, both involving balancing selection, to explain the variation in the *Mhc* genes. These are (1) heterozygote advantage and/or (2) frequency-dependent selection in response to parasites and pathogens (reviewed in Penn and Potts 1998, Hedrick 2002). There is not yet any consensus on which of these hypotheses is more important, although present evidence seems to lean towards some form of frequency-dependent selection (Sommer 2005, Hedrick 2006). Both mechanisms can explain why *Mhc* diversity is often high, even in species or populations were neutral markers indicate a loss of genetic variation due to random genetic drift (e.g. Aguilar *et al.* 2004, van Oosterhout *et al.* 2006).

It should be clear from the above that exactly how pathogens maintain a high level of *Mhc* diversity is still debated and the issue needs further investigation (Penn 2002, Zelano and Edwards 2002, Milinski 2006, Piertney and Oliver 2006). To clarify these issues, isolation of *Mhc* markers in non-model species is needed. This has until recently been hampered by interspecific variation in *Mhc* architecture. Since species vary considerably in the number of functional and non-functional *Mhc* genes, an important prerequisite to studying MHC diversity is to know how many duplications of *Mhc* genes are present in the species of interest, and whether or not these loci are expressed (Strand *et al.* 2007). Incomplete knowledge may lead to misleading conclusions, for instance if variation in pseudo-genes is associated with ecological factors. There is thus a need to understand how *Mhc* diversity is selected for and maintained in natural populations. Studies of associations between *Mhc* diversity, or MHC profile, and condition parameters (and mate choice) are frequent in the recent *Mhc* literature and include mammals, birds, and fish (Piertney and Oliver 2006). To better understand these interactions *Mhc* genes other than those coding for the classically studied MHC class II need to be targeted (Acevedo-Whitehouse and Cunningham 2006).

5.1.1 *Mhc* and conservation in mammals

The MHC was first discovered in humans in the 1950s in studies on skin graft rejection. The link to immunology was soon detected and at present more than 420 genes are known in this gene complex, of which 252 are expressed; about 70 of these are potentially associated with immunity (Beck and Trowsdale 1999). In humans the *Mhc* genes reside on chromosome 6 but may be regulated by genes located on other chromosomes (Reith and Mach 2001). It appears that *Mhc* structure and organization is quite similar in our close relatives, the great apes, with humans and chimpanzees sometimes sharing the same alleles. This trans-species polymorphism is a common observation in many mammalian studies (Klein *et al.* 1998, Garrigan and Hedrick 2003) and is explained by balancing selection maintaining variation for long periods. As a consequence, often the most similar *Mhc* sequence is not in the same species but in a related one (Hedrick 2006). In artiodactyls, balancing selection appears to have maintained allelic lineages for over 20 million years (Gutierrez-Espeleta *et al.* 2001).

Within-species genetic variation at *Mhc* loci can either be similar to that at neutral loci or, because of past balancing selection, exceed the neutral variation (the third possibility that neutral variation exceeds *Mhc* variation is to my knowledge never observed). Historical demographic events have been implicated to explain why Swedish beaver, *Castor fiber*, and moose, *Alces alces*, possess a low number of *Mhc* alleles (Ellegren *et al.* 1993, 1996, Mikko and Andersson 1995; see also Sommer 2005 for a list of similar examples, and Babik *et al.* 2005 for more on low *Mhc* diversity in Eurasian beavers). Bottlenecks and founder effects have according to this explanation been stronger than the power of selection in shaping current levels of *Mhc* variation. In these cases the reduced *Mhc* polymorphism is thus correlated with low genome-wide genetic variation (Hedrick 2002). African cheetahs, *Aconyx jubatus*, have been cited as the prime example in which low *Mhc* diversity correlates with a genome-wide loss of diversity, presumably due to a genetic bottleneck about 10 000 years ago (O'Brien *et al.* 1985). However, the details of this case have been debated. Another famous example is the Northern elephant seal, *Mirounga angustirostris*, which became almost extinct due to hunting about 100 years ago. This species is low in presumably neutral allozymes, mitochondrial DNA, mini- and microsatellite loci, as well as adaptive *Mhc* class II genetic variation (Hoelzel *et al.* 1999, Weber *et al.* 2004).

Other cases (reviewed by Sommer 2005) show that *Mhc* diversity may be maintained, at least for some time, despite the species being subject to loss of overall genetic variation. Although a direct link between pathogen-mediated population decline and low *Mhc* variation has been difficult to demonstrate in natural populations (Guiterrez-Espelata *et al.* 2001), recent studies have indicated that although *Mhc* allele numbers are low in many bottlenecked species, a high degree of divergence between alleles can still be observed. Moreover, genetic diversity at antigen-binding sites exceeds that at other *Mhc* codons in a range of threatened and fragmented species (Sommer 2005).

In summary, a few studies of mammals hint at the importance of *Mhc* variability in conservation (reviewed by Sommer 2005) although others indicate that species can persist, at least in the short term, despite being devoid of *Mhc* variability (Ellegren *et al.* 1993, 1996, Mikko and Andersson 1995). The importance of *Mhc* variability with respect to the severity of human impact is even less well studied. Theoretically one would expect a genotype-by-environment interaction whereby low variability might not lead to extinction when environmental conditions are benign, whereas adverse effects of low variability would become apparent under adverse environmental conditions due to human-induced changes such as pollution and habitat fragmentation.

5.1.2 *Mhc* and conservation in birds

Most of what is known of the genomic organization of *Mhc* in birds mostly comes from studies of the domestic chicken (Zoorob *et al.* 1993, Kaufman *et al.* 1999). Although more genomic information from other bird species is on the way, there is a need for specifically targeted studies of the comparative genomics of bird *Mhc*. The chicken *Mhc* gene family differs from mammalian *Mhc* by consisting of two independently assorting clusters of genes, the B and Y (formerly Rfp-Y) regions (Miller *et al.* 1996). Both these regions map to microchromosome number 16 in the chicken, and both contain *Mhc* class I and II genes (Miller *et al.* 2004). The B genes are polymorphic and expressed (Goto *et al.* 2002) and have been found to be correlated with resistance to several diseases in chickens (Kaufman 2000).

In chicken and some other birds there appear to be two expressed separate class II B genes (Freeman Gallant *et al.* 2002) but the number of both class I and II B genes may be manifold in other species (Westerdahl *et al.* 2000). Less is known about the Y genes. At least one *Mhc* class I Y locus (YF) is expressed and may be active in the immune function of the chicken (Hunt *et al.* 2006). However, to date it is not clear whether the *Mhc* class II Y (YLB) genes are functional in the chicken, as all the YLB loci mapped to date are apparently pseudo-genes (Shiina *et al.* 2006). The *Mhc* class II B (BLB) and YLB genes have only been characterized in chicken (e.g. Miller *et al.* 1996) and black grouse (Strand *et al.* 2007), but the ring-necked pheasant (Wittzell *et al.* 1995) and other birds also seem to have this division of *Mhc* class IIB genes. Studies of possible YLB genes will add to the understanding of the selection and evolution of *Mhc* genes in general. So far the *Mhc* class II studies of non-model bird species have, with the exception of our study on black grouse (Strand *et al.* 2007), focused on BLB or BLB-like genes (Table 5.1).

Table 5.1 Examples of MHC studies in non-model bird species.

Organism	Gene(s)	Finding	Reference
House sparrow, *Passer domesticus*	*Mhc* class I	Resistance to malaria	Bonneaud *et al.* 2004, 2006
Bobwhite quail, *Colinus virginianus*	B haplotypes	Polymorphism detected	Drake *et al.* 1999
Redwing blackbird, *Agelaius phoeniceus*	*Mhc* class II B and pseudo-genes	Polymorphims in functional gene	Edwards *et al.* 1998, 2000
Great snipe, *Gallinago media*	*Mhc* class II B	Polymorphisms	Ekblom *et al.* 2003
Savannah sparrow, *Passerculus sandwichensis*	*Mhc* class II B	Polymorphisms	Freeman-Gallant *et al.* 2002
Hawaiian honeycreepers (Drepanidinae)	*Mhc* class II B and pseudo-genes	Polymorphims in functional genes	Jarvi *et al.* 2002
New Zealand robins (Petroicidae)	*Mhc* class II B	Genes transcribed	Miller and Lambert 2004
Acrocephalus, warblers	*Mhc* class I	Polymorphisms in both inbred and outbred species	Westerdahl *et al.* 1999, Richardson and Westerdahl 2003
Bengalese finch, *Lonchura striata*	*Mhc* class II B	Presence of locus verified	Vincek *et al.* 1995

Several studies suggest that BLB genes are important in conservation. The Chatham Island black robin, *Petroica traversi*, found only, as the name indicates, on the Chatham Islands off New Zealand, is a highly inbred, endangered passerine with extremely low levels of genetic variation. Miller and Lambert (2004) investigated *Mhc* class II variation in both the black robin and its non-endangered relative, the South Island robin, *Petroica australis australis*. To test whether *Mhc* genes were under balancing selection they compared *Mhc* variation in the black robin with artificially bottlenecked populations of the South Island robin, and with their respective source populations. The black robin was monomorphic at the studied class II B loci, while both source and bottlenecked populations of South Island robin were found to have retained moderate levels of variation. Thus it was concluded that genetic drift must have outweighed balancing selection in the case of the black robin and consequently this species is extremely vulnerable to the introduction of new pathogens to the population.

The adaptive radiation and speciation of Hawaiian honeycreepers is a textbook example, but they currently face one the highest extinction rates in the world. The introduction of avian malaria to the Hawaiian islands is thought to be a major threat to extant honeycreepers. Jarvi *et al.* (2004) studied class II *Mhc* variation in four species of honeycreeper. Phylogenetic analyses revealed two clusters of genes and the authors found that variation in one cluster was high, with $dN > dS$ and levels of diversity similar to other studies of *Mhc* B genes in birds. The second cluster was nearly invariant, as in the studies of the Y genes in chicken and black grouse mentioned above. The presence of balancing selection was supported by transpecies polymorphisms and high dN/dS ratios at putative antigen-binding site codons. When comparing two species, mitochondrial DNA control region sequences were invariant in one species, but were highly variable in another. However, *Mhc* class II B variation appeared comparable. Thus, even though honeycreepers have been subjected to strong bottlenecks, it was concluded that balancing selection had been strong enough to maintain MHC variation.

The Galápagos Islands harbour the endemic Galápagos penguin, *Spheniscus mendiculus*, which is the only penguin that occurs on the equator. This species relies on food brought about by the nutrient-rich upwellings from the Humboldt stream and experiences severe population declines when ocean temperatures rise during so-called El Niño events, which occur irregularly. The reduced genetic diversity in this species are likely caused by the bottlenecks brought about by El Niño. Bollmer *et al.* (2007) characterized the amount of genetic variation at the *Mhc* in Galápagos penguins, and compared it with published data from other penguin species. They found that the Galápagos penguin had the lowest *Mhc* diversity of the eight penguin species studied. The authors explained an excess of non-synonymous mutations and a pattern of trans-specific evolution by suggesting that balancing selection may have been acting on the penguin *Mhc*. Thus this case mirrors the honeycreepers in that *Mhc* variation seemed to be upheld despite lost variation at neutral loci.

In studies from my own research group of a threatened lek-breeding wader, the great snipe *Gallinago media*, we found a high number of *Mhc* alleles (50 from 175 individuals; Ekblom *et al.* 2007). This, together with a higher rate of non-synonymous than synonymous substitutions in the peptide-binding sites, and high Tajima's *D* value in certain regions of the gene, suggests a history of balancing selection (Ekblom *et al.* 2008). Furthermore, genetic differentiation in the *Mhc* between two ecologically distinct distributional regions (Scandinavian mountain populations and Eastern European lowland populations) was present after statistically controlling for the effect of selectively neutral microsatellite variation (Fig. 5.2). This suggests that spatially varying selection is generating this structure and that this mechanism contributes to the balancing selection. *Mhc* variation in great snipe can thus be seen as a form of local adaptation to different environments. If this pattern is common, the implications for conservation are important. It suggests that local populations may be adapted to the local parasite fauna and that translocations of birds between populations may not do any good to their conservation status under certain circumstances.

Again, as is the case with mammals, studies on birds are equivocal on the conservation effects of lost *Mhc* variation. As in the case with the black robins reviewed above, species can persist despite having very little *Mhc* variation. In other case studies in which the study species perhaps have not been bottlenecked as severely, balancing selection seems to uphold MHC variation despite past and present population size reductions. Future studies are needed to resolve whether species can persist despite being reduced in *Mhc* variation. In theory it is just a matter of time before these *Mhc*-reduced species are hit by a new pathogen to which they cannot respond. If so, such taxa are doomed to extinction.

5.1.3 *Mhc* and conservation in reptiles and amphibians

Mhc structure and variation is poorly known in reptiles, but broad taxonomic studies have involved crocodiles and tuataras (see Edwards *et al.* 1995, Miller *et al.* 2005). There is to date no complete genomic mapping for any reptilian species. Similarly, very little information on *Mhc* variation and patterns of evolution are available for amphibians, a group known to be declining rapidly worldwide. Fungal diseases are most likely involved in these declines (Daszak *et al.* 1999, Pounds *et al.* 2006) and therefore information on *Mhc* could contribute to devising appropriate conservation strategies. *Mhc* class I and II has been characterized in two species of Urodela, in the axolotl *Ambystoma mexicana* (Sammut *et al.* 1997, 1999, Laurens *et al.* 2001, Richman *et al.* 2007) and class II in the tiger salamander *Ambystoma tigrinum* (Bos and DeWoody 2005), and in a single

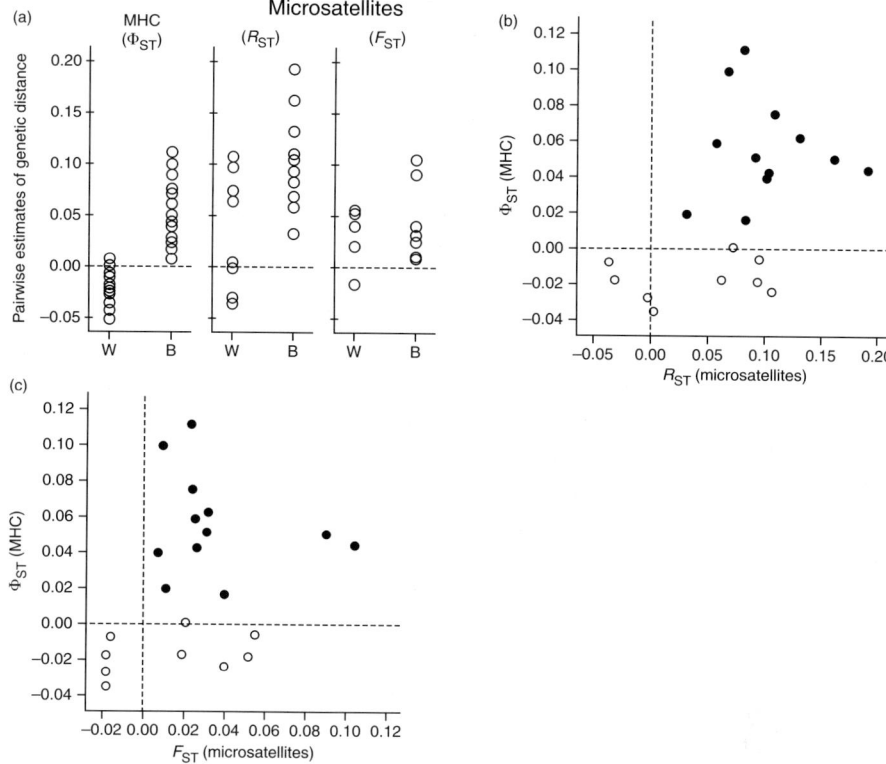

Figure 5.2 Point estimates of pairwise genetic distance (F_{ST}) between populations, within the same region (W) and in different regions (B), for *Mhc* class II genes and microsatellites. Different estimators of F_{ST} are used for the different markers (Φ_{ST} for *Mhc*; R_{ST} and Weir–Cockerham F_{ST} for microsatellites). For *Mhc*, estimates are larger for pairs of populations located in different regions than for estimates within region, and more so than expected from the corresponding patterns of (b, c) microsatellite pairwise estimates between regions (filled symbols) and within regions (open symbols) (from Ekblom *et al.* 2007, reprinted with permission from the publisher).

genus of anuran, the model taxon *Xenopus* (references in Ohta *et al.* 2006). Since the anuran model *Xenopus leavis* is a polyploid, anuran genome mapping has concentrated on the related diploid *Xenopus tropicalis*, in which a complete mapping of the entire *Mhc* exists (Ohta *et al.* 2006). *Mhc* genes appear to be evolutionarily conserved in *Xenopus* and all *Mhc* clusters are closely linked (Flajnik *et al.* 1999).

An early study compared mini- and microsatellite variation with *Mhc* in sand lizards, *Lacerta agilis*, and adders, *Vipera berus* (Madsen *et al.* 2000). In this study the authors were able to test whether smaller populations harboured less

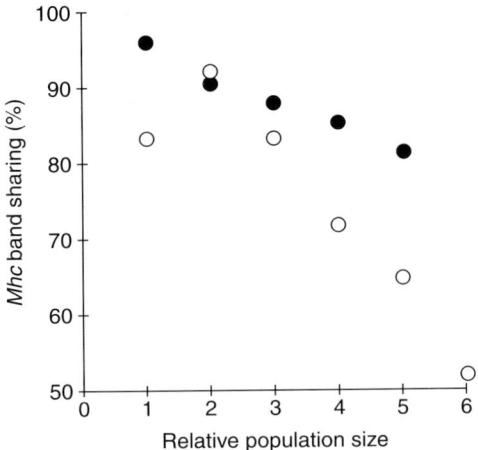

Figure 5.3 Mean *Mhc* class I band sharing and relative population size in six sand lizard populations (open circles) and five adder populations (filled circles) (from Madsen *et al.* 2000, reprinted with permission from the publisher).

genetic variation for each of the marker systems. It was found that this was only borne out in the case of *Mhc* (Fig. 5.3). Thus, it was argued that mini- and microsatellite techniques may provide ambiguous information concerning the relationship between population size and genetic variability. This is somewhat surprising since, due to balancing selection as argued above, a marker such as the *Mhc* may be the one expected to be deviant from a relationship between population size and genetic variability.

Some of the authors from the study cited above have also studied *Mhc* class I variation in isolated Hungarian populations of meadow vipers, *Vipera ursinii*, and compared them with larger Ukranian populations (Ujvari *et al.* 2002). Genetic variability at the class I loci was lower for Hungarian snakes than for Ukrainian populations. In Hungary, birth deformities, chromosomal abnormalities, and low juvenile survival was found, which strongly suggests that the Hungarian vipers are experiencing inbreeding depression. This study thus supports the notion that low *Mhc* variability is somehow tied to inbreeding depression. Unfortunately there is no information on overall genetic variability in these populations and therefore it is not possible to rule out the possibility that the observed inbreeding depression may be tied to an overall loss of genetic variation rather than being specifically connected with loss of MHC variability.

The variability of the peptide-binding region of *Mhc* class II in the fire-bellied toad *Bombina bombina*, which is of conservation concern in at least parts of its range, has been investigated and eight distinct alleles in 20 individuals were identified (Hauswaldt *et al.* 2007). All substitutions but one were non-synonymous, and many of the highly polymorphic sites corresponded with amino acid positions known to be involved in antigen binding. The level of *Mhc* variation found in fire-bellied toads was thus comparable with what has been found in other amphibians. Future studies are needed to resolve whether *Mhc* variation correlates with population size and is related to vulnerability to pathogens and extinction risk.

Similarly Babik *et al.* (2008) examined *Mhc* variation in Alpine newts *Mesotriton alpestris* from three allopatric population groups in Poland at the north-eastern margin of the distribution of this species. They found two putative expressed *Mhc* II loci with contrasting levels of variation. One locus exhibited low polymorphism. The other locus was highly polymorphic (37 alleles in 149 individuals), and showed evidence of balancing selection with populations varying substantially in allelic richness. The *Mhc* variation at this locus correlated with variation in microsatellites. The authors argued the observed regional differences could be explained by increased levels genetic drift with increasing distance from glacial refugia. This implies that selection and drift interplayed to produce the pattern of *Mhc* variation observed in marginal populations of the alpine newt and that marginal populations are more prone to extinction.

The number of studies on reptile and amphibian *Mhc* genes in a conservation context are too few to allow any firm conclusions. Technical advances in primer design (Hauswaldt *et al.* 2007) provide great promise for future studies of anuran non-model species. Such are highly relevant under the current decline in amphibians and especially relevant because the decline is most probably related to the spread of a pathogen (Berger *et al.* 1998, Daszak *et al.* 1999, 2003).

5.1.4 *Mhc* and conservation in fish

Mhc variation is relatively well studied in fish. In teleosts, MHC class I and II are not found on the same chromosome (Stet *et al.* 2003). By far the largest number of expressed class I and class II alleles are described for salmonids in which there is only one expressed class II locus (Grimholt *et al.* 2000). The fish model, the zebrafish *Danio rerio* and the three-spined stickleback *Gasterosteus aculeatus*, have been fully sequenced and in the three-spined stickleback *Mhc* genetics and ecology have been studied extensively (Milinski 2006). In sticklebacks there is more than one class II locus.

Genetic variation at eight microsatellite loci and sequence variation at exon 2 of the *Mhc* class II B genes in two wild populations of the Trinidadian guppy, *Poecilia reticulata*, were studied by van Oosterhout and coworkers (2006). They compared genetic variation in a small and isolated population upstream a system of rapids separating this population from a larger downstream population. Microsatellite diversity in the small population upstream was lower and the populations were genetically differentiated when considering microsatellites.

However, the two populations were not differentiated by *Mhc* and showed similar levels of allelic richness. The authors used computer simulations to suggest that the observed level of genetic variation in the two populations can be maintained with overdominant selection acting at three *Mhc* loci. This explanation requires that selection intensities varies among the populations and this is indeed what was found. Estimated selection intensities and parasite abundances suggested that large differences in selection intensity may exist between populations. Thus it is possible that high levels of *Mhc* diversity could be maintained in the small upstream population despite strong genetic drift.

Mhc studies on salmonids have been plentiful. There may be two reasons for this. First, there is a link between olfaction and *Mhc* and salmonids rely heavily on olfactory communication (Höglund 1961). MHC molecules are volatile and believed to be recognizable by olfaction (Wedekind and Füri 1987). Second, salmonids are of considerable economic importance in many fisheries and thus both pure and applied research resources have been avialbale for *Mhc* studies. As a result of this resarch, *Mhc* loci are used in the identification of harvesting stocks in the north-east Pacific (Withler *et al.* 1997, Beacham *et al.* 2001).

Patterns of population differentiation at neutral markers and *Mhc* genes have been studied in wild Atlantic salmon, *Salmo salar* (Landry and Bernatchez 2001). Variation at a *Mhc* class II B locus and microsatellites were compared among 14 samples from seven different rivers and seven subpopulations within a single river system covering a variety of habitats and different geographical scales. It was shown that balancing selection was acting on the sites involved in antigen presentation and thus could explain a high level of polymorphism within populations. The comparison of population structure at *Mhc* and microsatellites on large geographical scales revealed a correlation between patterns of differentiation despite important differences in habitat type among populations. This indicated that genetic drift and migration have been more important than selection in shaping population differentiation at the *Mhc* locus. On the other hand there were strong discrepancies between patterns of population differentiation among the two types of marker within rivers, which suggested a role of selection in shaping population structure at this scale. Taken together these results suggest that both selection and drift are influencing *Mhc* gene diversity in wild Atlantic salmon. This is a very similar result to the one found in the great snipe study reviewed above and confirms that translocations as a conservation measure should be considered with caution.

Studies of *Mhc* genetic variation among populations of chinook salmon, *Onchorynchus tshawytscha*, at three class I loci and one class II locus showed that populations from different river drainages were differentiated (Miller *et al.* 1997). As in Atlantic salmon it appears as though *Mhc* variation has been shaped by a combination of selection and genetic drift. The force of genetic drift has been influenced by repeated bottlenecks and isolation by distance in separate glacial refugia. Again, these populations seem at least partly adapted to local parasitic faunas and any conservation measures should be taken with this in mind.

The conservation status of another salmonid fish, the brown trout, *Salmo trutta*, varies across its distribution. Many populations are threatened by various types of human activity, like environmental degradation, harvesting, and development of hydroelectric dams (Laikre and Ryman 1996). Campos *et al.* (2006) studied levels and distribution of genetic variation in nine isolated populations of Brown trout in northern Spain and tested the importance of preservation of genetic variability for the survival of a set of isolated populations from the same river drainage system. They screened genetic variation at three different markers: mitochondrial DNA, microsatellites, and the *Mhc* class II locus. Genetic variation was similar at *Mhc* loci and microsatellites: populations polymorphic for microsatellite loci were also polymorphic at the *Mhc* loci (Fig. 5.4). They also observed high levels of differentiation among populations. Thus, in this case genetic drift seemed to have eroded the effect of balancing selection and was seen as the predominant evolutionary force shaping genetic variation in the smaller populations.

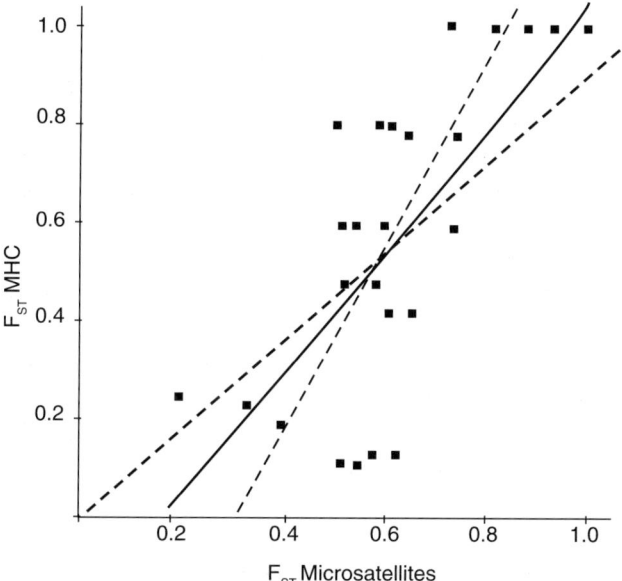

Figure 5.4 Regression of pairwise F_{ST} for MHC and microsatellite loci in Spanish brown trout populations. Dotted lines indicate 95% confidence limits (Campos *et al.* 2006, reprinted with permission from the publisher).

It has been noted that local brown trout populations may be highly differentiated and adapted to their local stream and thus it may be important to take genetic variation between populations into account in conservation programmes (Laikre 1999). As MHC molecules have the potential to coevolve in response to selection pressures imposed by local parasite faunas, *Mhc* variability may be of a special concern in maintaining local adaptations. Thus it may be argued that the basic unit for management and conservation of brown trout (or any other organism) are the local populations (Laikre 1999). If genetic variation and population structure at the *Mhc* and any other locus is mainly driven by neutral processes then local adaptation is prohibited, but spatially varying selection may lead to local adaptation, especially if migration is limited. If this is the case, admixture of local populations may lead to outbreeding depression as has recently been shown in a Swedish brown trout population (Grahn and Forsberg 2008). In the River Dalälven in Sweden a hydroelectric power plant has hindered migration to previously used spawning grounds since 1915 and artificial breeding and stocking have been provided as a substitute. It is likely that local populations within the river have become admixed during this process and experimental data show that in this admixed population *Mhc* homozygous males have an advantage in spawning competition and in production of young. The explanation for this surprising and counterintuitive result is that the *Mhc* diversity in the artificially admixed population is too high and above the optimal level. Reasons for why high variation may lead to fitness loss include autoimmune responses (Nowak *et al.* 1992, Reusch, *et al.* 2001) and loss of local adaptation to prevailing conditions.

As in the case with other animals the evidence for *Mhc* polymorphism being maintained by balancing selection is somewhat ambiguous in fish but it seems clear that *Mhc* variation in many populations is indeed maintained by balancing selection imposed by local parasitic faunas. This emphasizes what was hinted by studies on other vertebrates: local adaptation in *Mhc* is prevalent and admixture of previously locally adapted populations may have adverse effects, as in the case of Swedish brown trout.

5.1.5 Summary: *Mhc* and immunogenetics in conservation

Because of the role of the MHC in the immune defence of vertebrates, *Mhc* variability is arguably important for the viability of natural populations. As reviewed above, many studies have shown that populations exhibiting low levels of variability at the *Mhc* or with certain haplotypes are susceptible to diseases and therefore prone to extinction (see also O'Brien *et al.* 1985, Paterson *et al.* 1998, Langefors *et al.* 2001, Arkush *et al.* 2002). However, other studies have presented evidence that populations with no or low variability at *Mhc* loci are still persisting (Slade and McCallum 1992, Ellegren *et al.* 1993, Seddon and Baverstock

1999, Hedrick *et al.* 2000, Miller and Lambert 2004, Weber *et al.* 2004). The role and importance of *Mhc* genes in conservation are thus debated. There has been a strong focus on *Mhc* class II genes in many immunogenetic studies with a conservation focus. While these studies have provided much insight into disease resistance in wild and threatened populations, is clear that conservation immunogenetic studies will benefit by including more immune genes and loci in the future (Acevedo-Whitehouse and Cunningham 2006).

An example of such a study is one on Danish brown trout populations (Jensen *et al.* 2008). These authors used eight neutral microsatellite loci and two microsatellite loci embedded in the sequence encoding the protein TAP (which stands for transporter associated with antigen processing) to study temporal and geographic differentiation. *Tap* genes encode molecules that associate with MHC class I molecules when foreign peptides are transported across the membrane of the endoplasmatic reticulum and thus are important in launching an immune response to intracellular parasites. Thus the genetic variation at these loci could be influenced by parasite- and pathogen-driven selection. The observed neutral genetic variation suggested that population structure was temporally unstable within regions, although stable over time among regions. Statistical tests designed to detect selective sweeps found evidence of selection at the two *Tap* markers, indicating both a regional and microgeographical effect. Moreover, signals of divergent selection among temporal samples within localities suggest that selection also might fluctuate at a temporal scale. These results suggest that immune genes other than the classical *Mhc* classes I and II might be subject to selection and warrant further studies of functional polymorphism of such genes in natural populations.

5.2 Other candidate genes relevant for conservation

Mhc genes have been by far the most commonly studied candidate genes in the context of conservation. Other genes have been less studied, partly because relevant genomic information has been scarce in non-model species. As technical advances proceed there is no reason for not including other ecologically important genes when studying threatened and endangered species. Below I briefly review a few genes which have been studied in non-model species in an evolutionary ecology framework. Such studies obviously have a bearing on conservation issues (Segelbacher and Höglund 2008).

5.2.1 Pigmentation genes: *mc1r*

The study of animal pigmentation has a long history in ecological genetics (Hoekstra 2006). The classical studies of banding patterns in *Cepaea* snails and industrial melanism in the peppered moth *Biston betularia* serve as only two examples of how the study of evolutionary genetics of coloration have played an important role in understanding how populations may adapt to local differences in selective regimes (in these two cases ultimately driven by visual predators). It is clear that pigmentation has a strong genetic component and that populations quickly can adapt to local conditions (Majerus 1998). Pigmentation genes should therefore be very relevant in a conservation context.

Although there are several types of animal pigments the most studied and well-known system is that of melanin-based pigmentation. Melanin is produced by specialized cells, so-called melanocytes. Melanin production, or melanogenesis, in vertebrates is a complex process that includes the inception, migration, and regulation of melanocytes (Jackson 1994). Melanocytes can synthesize either eumelanin or phaeomelanin, or produce no pigment at all. Increased eumelanin synthesis leads to darker skin, hair, or feathers, increased production of phaeomelanin produces red or brown phenotypes, and no melanin synthesis results in albinism (Fig. 5.5).

The physiological pathways and the genes involved in melanin-based pigmentation have recently become quite well established. In mammals the best-known pathway is the one mediated by the cell-surface protein MC1R (melanocortin 1 receptor or α melanocyte-stimulating hormone receptor). Here circulating levels of the agonist α melanocyte-stimulating hormone (αMSH) activate MC1R which triggers the production of a messenger molecule cAMP which activates a complex pathway involving tyrosinase (Tyr) and tyrosinase-related protein 1 (Tyrp1), ultimately leading to synthesis of eumelanin. If Agouti, which is the inverse antagonist of αMSH, binds to MC1R, the outcome is no synthesis of melanin or phaeomelanin. Melanin synthesis is believed to follow a similar pathway in other animals but the details are less well known (Mundy 2006, Hoekstra 2006).

The *mc1r* gene is a short gene (the single exon extends approximately 1000 bp) expressed in melanocytes in skin and developing feather buds or analogue tissues in vertebrates. In humans it is known that mutations on *mc1r* are often correlated with phenotypic variation such as red hair and light skin (Makova and Norton 2005). Studies linking phenotypic variation with sequence polymorphisms have also been published in both domesticated (e.g. Kerje *et al.* 2003, Våge *et al.* 2005) and wild (reviews by Mundy 2006, Hoekstra 2006) animals. Recently, *mc1r* evolution have been shown to evolve faster in lineages of galliforms that show more plumage dimorphism. This is probably due to varying intensities of sexual selection (Nadeau *et al.* 2007a).

As in the case with immune genes there is a strong focus on a single candidate gene in the study of pigmentation genes. In a recent review of pigmentation mutations segregating in wild vertebrate populations Hoekstra (2006) listed 14 studies of which 12 were on *mc1r*. Clearly there is a need for studies of more

Other candidate genes 97

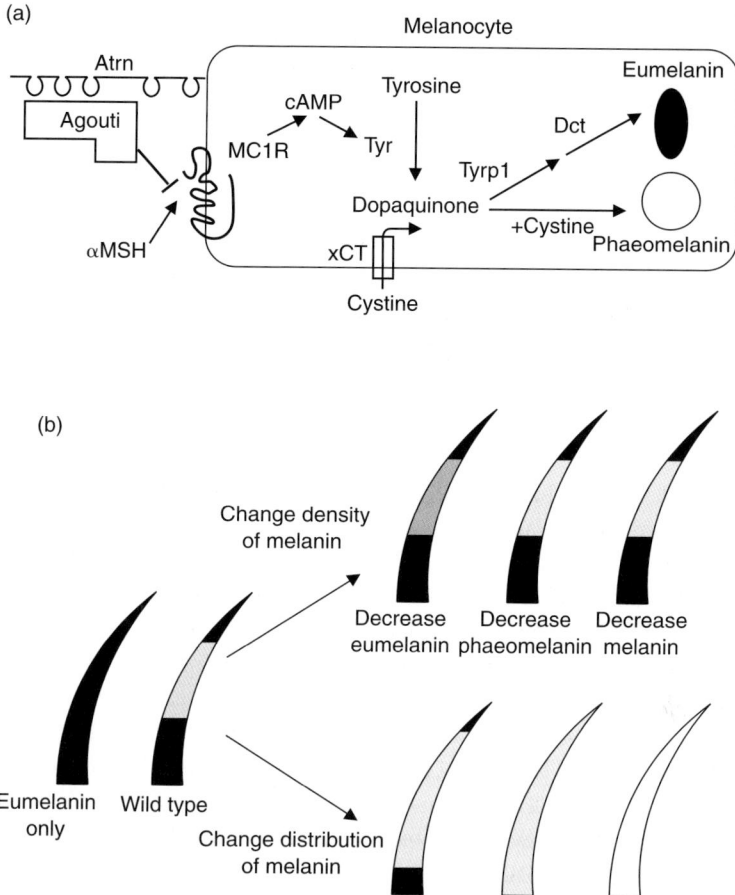

Figure 5.5 Schematic representation of the pathways regulating mammalian melanogenesis and phenotypic effects on individual hair pigment and pattern. (a) The binding of circulating α melanocyte-stimulating hormone (αMSH) to MC1R activates the synthesis of the enzyme tyrosinase (Tyr) via cAMP. Within the cell, tyrosine is oxidized to dopaquinone, a reaction catalysed by Tyr. cAMP affects the enzymatic activity of Tyr as well as the eumelanic-specific enzymes, tyrosinase-related protein 1 (Tyrp1) and dopachrome tautomerase (Dct). When all three of these enzymes are working, eumelanin (brown to black pigment) is deposited in melanosomes. However, when Agouti, the inverse agonist of MC1R, binds to MC1R with the aid of the extracellular protein Atrn, intracellular cAMP levels are repressed and this leads to production of phaeomelanin (yellow to red pigment) which is also dependent on the incorporation of cystine, whose uptake is at least partially regulated by xCT (a protein regulating cystine uptake in melanocytes; in mice it is a gene product of the Slc7a11 locus). (b) Illustration of how overall coat colour in mammals is determined by the density and distribution of melanin on individual hairs. Pigmentation on individual hairs ranges from fully pigmented with dark eumelanin to complete absence of pigment resulting in albino hairs (from Hoekstra 2006, reprinted with permission from the publisher).

loci involved in melanin synthesis and of other pigments. For example, studies in Japanese quail, *Coturnix japonica*, have shown that there is an association between a single-nucleotide substitution in the gene encoding Tyrp1 and plumage colour (Nadeau *et al.* 2007b).

To date I am aware of no studies directly relating variation in any pigmentation gene to conservation issues. However, studies of pigmentation in a conservation context would probably challenge the view that preserving genetic diversity *per se* is all that matters in conservation genetics. Since pigmentation is often strongly related to the ecological background of the organism there is often a match between the environment and the most optimal phenotype (Hoekstra *et al.* 2003, 2005). As in the case of *Mhc*, transplantation to boost numbers or genetic variation could introduce alleles that have negative consequences.

5.2.2 Photoperiodism: *Clock* and other genes

Several recent studies have pointed out the circadian clock (i.e. synchronization of an organism to daily rhythms; Bell-Pedersen *et al.* 2005) as one of the aspects of animal behaviour best characterized at the molecular level (Fidler and Gwinner 2003, Johnsen *et al.* 2007). For example, in vertebrates the gene *Clock* encodes a protein that heterodimerizes with a second protein, BMAL1, to produce a transcription-activating complex which is important in the molecular control of vertebrate circadian rhythms (Panda *et al.* 2002). In humans, a single nucleotide polymorphism in *Clock* correlates with variation in sleeping-time preferences (Mishima *et al.* 2005). Not only circadian rhythms appear influenced by *Clock*; there is evidence that *Clock* polymorphisms are associated with differences in spawning times in rainbow trout, *Oncorynchus mykiss* (Leder *et al.* 2006).

Johnsen *et al.* (2007) studied allelic variation in a region of the avian *Clock* which encodes a polyglutamine repeat (*Clk*polyQcds), in two species of passerine birds, the bluethroat, *Luscinia svecica*, which is a migrant, and the non-migratory blue tit, *Cyanistes caeruleus*. Multiple *Clk*polyQcds alleles were found within populations of both species. When testing for population differentiation they found that observed allele frequency variation among populations at the *Clk*polyQcds and at neutral microsatellite loci could not be explained by the same underlying demographic processes in blue tits. In this species allelic variation in the *Clk*polyQcds showed evidence of being maintained by selection for microevolutionary adaptation to differences in photoperiod. This could not be detected among bluethroats, possible because of low statistical power due to small sample sizes.

The allelic variation in this case was found in polyglutamine repeats in the coding region of the gene. As pointed out by Johnsen and coworkers, it has been hypothesized that the relatively high mutation rates of repeat sequences may account for rapid morphological evolution among mammals (Fondon and Garner 2004) and they suggest further that the potential of coding region repeats for rapid evolution might be selected for, as it provides plasticity in the face of fluctuating selective pressures (see also Wren *et al.* 2000). They also proposed that investigating changing frequencies of allelic variants of genes encoding circadian clock components may warrant attention in the context of adaptation to rapid climate change. When climate changes, many parameters related to biorhythms are predicted to change accordingly. So if, for example, passerine bird populations are adapted to respond to changes in photoperiodism to time their maximum reproductive output with a phenological peak in food abundance, such populations either have to respond genetically or face extinction (Dias and Blondel 1997).

In plants there has been a quest to find the genes involved in the regulation of ecologically important traits such as flowering time, seed set, bud set, and annual differences in growth. Like circadian and phenological rhythms in animals, flowering time is an important life-history trait that coordinates the life cycle with local environmental conditions (Roux *et al.* 2006). The genetic basis of flowering time in plants have been studied extensively in the model species *Arabidopsis thaliana*. Such work has revealed a complex network of genes involved in flowering-time regulation (Fig. 5.6). There appears to be four major pathways and many potential candidate genes (Bernier and Périlleux 2005).

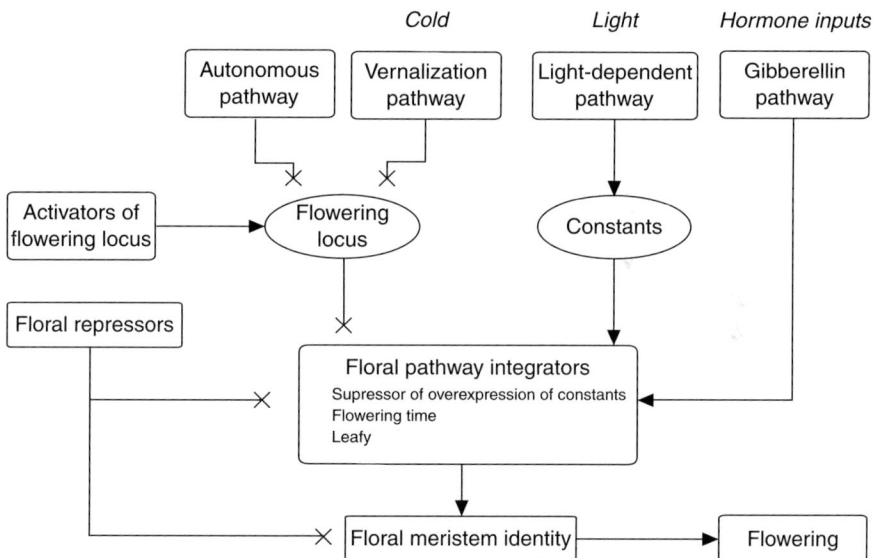

Figure 5.6 Simplified overall network of flowering-time regulation (from Roux *et al.* 2006, reprinted with permission from the publisher).

It is predicted that flowering time should correlate with latitudinal differences among populations. However, no clear patterns have been found (Stinchcombe *et al.* 2004, Shindo *et al.* 2005). It was suggested that geographical trends may be masked by other selective regimes than photoperiod that may vary locally (Roux *et al.* 2006). Despite the inconsistencies among studies it seems clear that photoreceptor genes are major agents of natural variation in *Arabidopsis* flowering and growth response as shown by genome-wide scans of association of 65 loci with latitude (Balasubramanian *et al.* 2006). In this study the most associated locus across 163 strains was *Phyc* (phytochrome C photoreceptor), suggesting that *Phyc* is under diversifying selection.

A major challenge is to transfer the results in *Arabidopsis* to other plants. In rice, genus *Oryza*, two independent floral pathways have been detected: one is mediated by *Hdf1* (heading date 1), which is an orthologue of the *Arabidopsis* gene *CO* (Constans), and the other is *Ehd1* (early heading date 1), an orthologue of *FT* (flowering time locus). Both *CO* and *FT* play important roles in the timing of flowering in *Arabidopsis*. Similarly, indel variation in a *CO* orthologue in *Brassica nigra* has been shown to be associated with variation in flowering time in this species (Österberg *et al.* 2002), and an *FT* homolog has been implicated in controlling growth rhythm in conifers (Gyllenstrand *et al.* 2007). Likewise, nucleotide polymorphisms in the *phytochrome B2* locus in aspen *Populus tremula* have have been found to be associated with the timing of bud set (Ingvarsson *et al.* 2008).

However, there are also important differences between *Arabidopsis* and wild species. Slotte and coworkers (2007) used gene-expression differences between pairs of early- and late-flowering *Capsella bursa-pastoris* ecotypes and compared their responses to changes in temperature (vernalization). Using *Arabidopsis* microarrays they found differences among the ecotypes. In contrast, in *Arabidopsis* FLC (flowering time locus C) was not differentially expressed prior to vernalization and the gibberellin and photoperiodic pathways appeared similar.

The picture that seems to emerge is that photoperiodism is evolutionarily rather conserved among plants. Thus there is scope for using a candidate gene approach to studying and preserving genetic diversity in threatened plant populations. However, as shown by the complexity among biological pathways such an approach is not without complications.

5.3 Conclusions

It is a conservation genetic paradigm that genetic variation is a prerequisite for any population's ability to adapt to a changing environment. Since small and fragmented populations are signified by low levels of genetic variation it follows that they are thus less able to adapt when conditions change. Population fragmentation and isolation thus have extremely detrimental effects on both the fitness and viability of extant populations, and also the evolutionary potential of species (see papers in Ferrière *et al.* 2004). This line of reasoning may lead to the conclusion that all that matters in conservation genetics is to preserve genetic variation and the more the better. However, as has been argued in this chapter, studies on ecologically relevant candidate genes to some extent challenge this view. Of course conservation genetics should still focus on the preservation of genetic variation and on detecting the processes that are important in preserving natural populations of threatened species, but preservation of genetic diversity must be done with knowledge and caution. Thus it is important to understand local adaptation, which is the topic of the next chapter.

Most studies that have attempted to monitor genetic diversity within and among threatened populations have used so-called neutral genetic markers to quantify variation (McKay and Latta 2002, Sommer 2005). As argued previously in this volume these markers are excellent for estimating effective population size, migration rates, and other population genetic processes since, on the whole, they are not affected by selection and hence genetic variation is mainly determined by genetic drift. However, it is now questioned whether neutral genetic variation is a suitable proxy for the ecologically meaningful genetic variation required to maintain populations as viable entities capable of adapting to habitat and environmental change (e.g. Madsen *et al.* 2000, Hedrick 2001). A recent review used the following citation from Frankham (1999) to illustrate the present situation on how to study genetic variation in natural populations using neutral markers, quantitative trait loci, and ecological traits: 'A major unresolved issue [in conservation] is the relationship between molecular measures of genetic diversity and quantitative genetic variation' (McKay and Latta 2002). Studies of candidate genes are bridges to understanding local adaptation. As has been discussed in this chapter, selection may both maintain genetic variation, through balancing selection, and erode it, through purifying and directional selection.

6 *Local adaptation*

The candidate gene approach reviewed in the previous chapter relies on the basic assumption that under certain circumstances particular genetic variants are favoured over others. This suggests that it should be possible to identify selective pressures and the genes that are favoured in natural populations; that is, it should be possible to find adaptations. Indeed, such studies have a long tradition in evolutionary biology (Rose and Lauder 1996, Banta *et al.* 2007). This chapter will cover evidence for local adaptation in natural populations. It is a paradigm in conservation genetics that when genetic variation is present populations can adapt to new circumstances but that this ability is hampered if genetic variation is lost or reduced and this is arguably the ultimate threat to endangered populations (Frankham *et al.* 2002, Latta 2008). As we have seen in previous chapters, human-induced changes to the environment tend to induce population structure among previously continuously distributed species and leave them to live in more and more fragmented habitats (see papers in Smith and Bernatchez 2008). These processes also have the consequence that effective population sizes become smaller and as fragments are lost populations tend to become more and more isolated. Thus migration is mitigated and genetic differentiation among populations increases. All this has consequences for the survival of threatened species and to boost genetic variation among small and isolated fragments genetic restoration brought about by transplantation of individuals is sometimes considered as a practical conservation action and is in some cases also performed (Westemeier *et al.* 1998, Madsen *et al.* 1999). As argued in previous chapters these concerns and actions are taken to counteract the deleterious of effects of genetic drift and inbreeding.

The deleterious effects on offspring due to inbreeding decrease as the genetic distance between parents increases but the increased offspring fitness is predicted to level off as the genetic distance between individuals becomes too large (Fenster and Galloway 2000). The extreme example of this is the reduced fitness of hybrid offspring when species are crossed. Even within species it is sometimes observed that hybrids between various subspecies may show reduced fitness

(Hewitt *et al.* 1987, Rubidge *et al.* 2001). Such cases of outbreeding depression suggest that there is an optimal level of inbreeding (Bateson 1983).

In subdivided populations, it is inevitable that local populations experience localized selection pressures. The reasons could be plentiful but local populations may respond to differences in factors such as altitude, light regime, soil conditions, predation pressure, and interspecific competition. In fact, any two local populations are much more likely to experience different than similar conditions. These different conditions may thus lead to local adaptations. Documented cases of local adaptation—that is, genetic changes in a local population in response to a local selection pressure—are plentiful in the literature (and briefly reviewed below). The importance of outbreeding depression and local adaptation are of concern in conservation biology as the occurrence of each would question whether artificial gene flow between isolated populations in the form transplantations is always a well-founded conservation strategy.

The issue of whether natural populations are more influenced by selection of drift has been a long-standing issue in evolutionary biology ever since the work of Kimura (1968). Are quantitative characters in natural populations shaped by natural selection or driven mainly by neutral processes (e.g. Smith *et al.* 1997)? This question can be addressed using so-called Q_{ST} versus F_{ST} comparisons. Such investigations compare levels of divergence among populations in quantitative traits with that of neutral characters (e.g. Merilä and Crnokrak 2001) and can thus be used to say something about the role of natural selection in natural populations. The rationale is that population differentiation observed in a neutral character would provide a baseline of population differentiation mediated through genetic drift, whereas the differentiation observed in a quantitative character could be shaped by both drift and selection and hence may be different from what is predicted from neutral loci (Spitze 1993).

6.1 Evidence of local adaptation

Evidence of adaptation to local environmental conditions is so plentiful it is hard to make a fair review of all the relevant studies. Searching the PubMed Internet database in June 2008 for 'evidence of adaptation to local environmental conditions' yielded 102 hits. What is mentioned in the following is thus not a comprehensive list but rather a few examples illustrating that local adaptation is a rule rather than an exception in natural populations. The issue in conservation biology is whether endangered populations have lost so much genetic variation that they have lost their adaptability.

There are two main methods used to study local adaptation in nature. The first is the direct, or so-called allochronic, method. In this approach changes over time within a population are followed. Usually the heritability of a trait suspected to be adapted to local conditions is determined and the fitness of phenotypes is correlated with changes in the environment. The studies on adaptive responses of body and beak morphology to varying conditions in Darwin's finches of the genus *Geospiza* (Grant and Grant 2002, 2007) and the adaptive changes of life-history traits and secondary sexual characters in guppies, *Poecilia reticulata* (Reznick and Bryga 1987), are examples of studies using the allochronic method.

The alternative to the direct method is the indirect, or synchronic, method. Here divergences between contemporary populations that share a common ancestor are studied and environment–phenotype correlations among study organisms in contrasting environments are determined. These correlations may then used to test the performance of individuals under experimental conditions, for example by determining the fitness of phenotypes in different environments by reciprocal transplantation experiments. The studies of blue tits, *Cyanistes caeruleus*, by Blondel and coworkers (1993) and on Caribbean *Anolis* lizards by Losos (1990) are examples of the synchronic method of detecting adaptations.

Reznick and Travis (1996) listed five examples of classic empirical studies of adaptation, as follows.

1. The case of industrial melanism in the peppered moth *Biston betularia* and the rise in frequency of the melanic morph. This was shown to be tied to increased levels industrial pollution which made the melanic morph match with the background and thus made them less visible to visual predators (Kettlewell 1955, 1956, 1958). More recent studies of this system have documented a fall in the frequency of the melanistic morph as industrial pollution has become less severe (Majerus 1998).

2. The banding patterns of *Cepaea* snails and its relation to background and the impact by visual bird predators (Cain and Sheppard 1950, 1954).

3. The case of *Peromyscus* mice polymorphic pelage colour and background matching which, like the first two cases, is related to bird predation (Dice 1947).

4. Heavy metal tolerance in *Agrostis tenuis*. Plants growing near mines with soils contaminated by heavy metals are less susceptible to heavy metal contamination than plants from uncontaminated soils (McNeilly 1968, McNeilly and Bradshaw 1968).

5. Chromosome inversion patterns in *Drosophila* in which the frequency of the inversion changes with altitude and season (Dobzhansky 1948).

Studies of plants have played a major role in increasing the understanding of local adaptation and there are many examples reporting results consistent with both large- and small-scale adaptation to local conditions. For example, reciprocal transplant experiments of the narrow-leaf plantain *Plantago lanceolata* showed that survival was best in plants originating from the 'native site' (van Tienderen and van der Toorn 1991). In this and many other studies the results may be confounded by geographic distance and clinal variation, possibly explained by genetic drift. However, a study of the shrub *Lotus scoparius* found that genetic distance and ecological similarity between the source and transplant population were stronger determinants of plant success than geographic distance (Montalvo and Ellstrand 2000). This strongly suggests that local adaptation produced by natural selection rather than genetic drift is the underlying cause of this result.

Local adaptation may be extremely fine scaled in plants. In the perennial herb *Gypsophila fastigiata* on the Baltic island of Öland, only approximately 2% of the total allozyme diversity was explained by differentiation between sites tens of kilometres apart. However, differentiation at the *Pgi-2* locus was significant at scales of only 10 m and was associated with habitat differences and differences in individual reproductive success (Lönn *et al.* 1996). This may be explained by differential selection due to microhabitat differences in the soil and suggests that differential selection may contribute to local fine-scale structuring despite extensive gene flow in an outcrossing species.

Evolution was long regarded as an exceedingly slow process. However, rapid responses to local selection and rapid responses to habitat changes seem possible in many organisms (Stockwell *et al.* 2003). Evolutionary changes can occur within decades or even shorter time spans (Fig. 6.1). As an example, studies of adaptations to temperature changes in the pitcher-plant mosquito *Wyemyia smithii* shows that the genetically determined critical photoperiod response (defined as the number of hours of light per day that initiates or maintains diapause in 50% of a sample population and averts or terminates diapause in the other 50%) has been significantly advanced throughout the altitudinal range of the studied populations since 1972 when measurements were first begun and even are detectable in as short a time period as 5 years (Bradshaw and Holzapfel 2001). This result is best understood as an evolutionary response to a warmer climate.

The Scandinavian peninsula encompasses several well-studied environmental gradients. First, the habitat is generally cooler and the growth season shorter towards the north. Second, running from west to east acidity (pH) declines. Generally speaking, habitats are more acidic in the west than in the east (Fig. 6.2). In series of papers Anssi Laurila and Juha Merilä and their colleagues studied adaptation in common frogs, *Rana temporaria*, and moor frogs, *Rana arvalis*, and in particular the larval life-history traits along these two gradients. Among many things they have shown that developmental rates (larval growth) in common frogs in the field varied extensively among different ponds. In contrast, development rates in the laboratory increased linearly with increasing latitude. Thus these results suggest that there is a genetic capacity for faster development

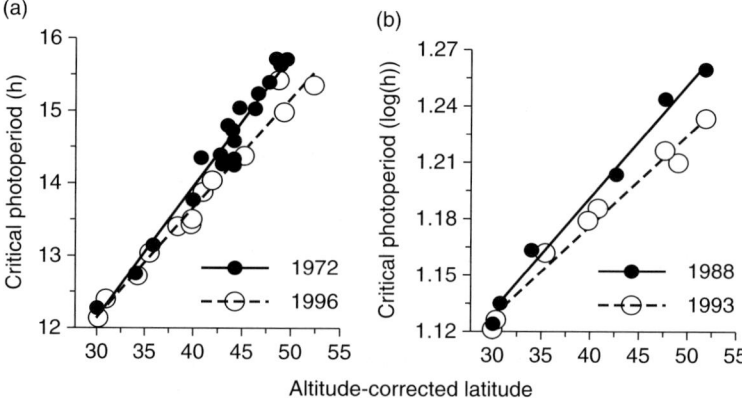

Figure 6.1 Shifts towards shorter critical photoperiods increase with latitude in the pitcher-plant mosquito *Wyemyia smithii*. This indicates selection for more southern phenotypes with increasing latitude (from Bradshaw and Holzapfel 2001, reprinted with permission from the publisher).

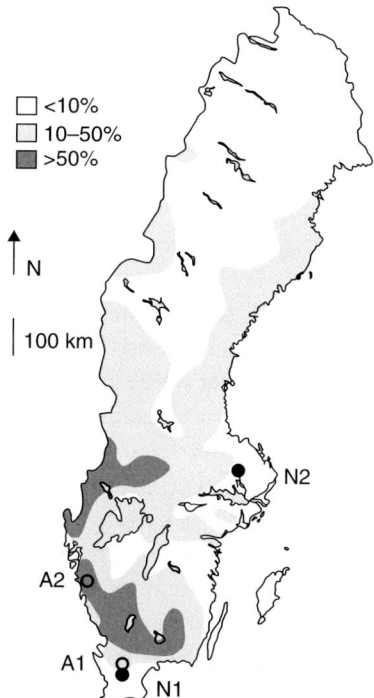

Figure 6.2 Map of Sweden showing the percentage of acidified lakes in Sweden in 1990 and locations of two acid (A1, A2) and two neutral (N1, N2) study populations (from Räsänen *et al.* 2003a, reprinted with permission from the publisher).

in the north. This is an example of countergradient variation were the realized trait in the field (in this case development time) is masked by a harsher environment and/or relaxed predator regimes (Laugen *et al.* 2003). In moor frogs they have shown that populations originating from acidic environments were more tolerant to acid environments during development (Räsänen *et al.* 2003a). The mechanism appears to be mediated via the female in that females from more acidic environments produce eggs that have a more protective gelatinous egg capsule surrounding the embryo (Räsenen *et al.* 2003b).

In my own research group we have used experimental studies to approach whether genetic diversity may affect adaptability by studying the extent of local adaptation in two disjunct distribution areas of the natterjack toad, *Bufo calamita*, in southern Sweden (B. Rogell *et al.*, unpublished results). In Sweden natterjack toads occur in two regions that differ dramatically in habitat. On the Swedish west coast in Bohuslän, several isolated natterjack toad populations occur on rocky off-shore islands . These islands are influenced by both saltwater influx and desiccation risk as their breeding ponds are shallow and often situated very close to the sea. Both these factors constitute selection pressures that are weaker in the other area of Sweden inhabited by natterjack toads: in southern Sweden, in the province Skåne, natterjack toad populations are found in deeper and more permanent ponds that are vegetated. We therefore, hypothesized that the Bohuslän populations should be locally adapted to elevated salinity levels and increased dissecation risk imposed by evaporating ponds. Traits for such adaptations could be increased salinity tolerance and higher developmental rates.

We raised tadpoles from several populations from both Bohuslän and Skåne in the laboratory in a common garden experiment with three levels of salinity in each of two temperature treatments. We recorded the number of days until and weight at metamorphosis and the percentage of surviving tadpoles in the treatments.

We had previously shown toads from larger populations in Bohuslän harbour more genetic variation than smaller populations in the same area (Chapter 2). We thus predicted that, because of their higher adaptive potential, the most genetically diverse populations would have adapted better to the selection pressure. Our results in this experiment confirmed this hypothesis: the larval development of genetically impoverished populations as compared with genetically more diverse populations took longer to complete their metamorphosis, indicating that low genetic diversity limits the adaptation to desiccation risk. Thus our results show that neutral genetic diversity predict how local populations may respond to local selection in that low genetic diversity hampered the adaptive response to dissecation risk.

We further found that tadpoles from both Bohuslän and Skåne were most severely affected by low temperature and high salinity. As predicted, the

Bohuslän populations were found to be locally adapted to desiccating ponds, but at the same time they showed higher mortality and were more severely impacted by elevated salinity than the Skåne toads. Our explanation for these apparently contradictory results are that we found a negative genetic correlation between fast larval development and salinity tolerance (Hofman 2007). It appears as though the toads are facing a trade-off between short development time and salinity tolerance. We hypothesize that the former outweighs the latter, resulting in the Bohuslän survival pattern not matching with the expectation of local adaptation to elevated salinity.

If local adaptation is important there are several issues of relevance for conservation biology. First, if locally coadapted gene complexes reside in small and endangered populations, transplantation or reintroductions may be bad conservation strategies since they make break up local adaptations or introduce suboptimal genotypes to localities to which they are not adapted. On the other hand, if adaptation is a common phenomenon and possible even in small populations within short time periods, concerns about breaking up coadapted gene complexes are superficial as they would rapidly be replaced not by the same but by new optimal phenotypes. These questions are likely to be answered by studies of the residing threatened population. In cases where the genetic variability and the population size are so low that genetic drift is the sole force shaping the genetic diversity of the population and in which there are clear signs of inbreeding depression, genetic rescue projects may be called for. These cases are most likely restricted to such when the effective population size is very low (in the order of or below 10 individuals). In other cases I would advise caution against transplantations and reintroduction programmes. Conservationists should strive to reach population sizes that ensure the future adaptability of the reintroduced population. Based on theoretical arguments, Lande and Shannon (1996) noted that 'evolutionary biologists and conservation geneticists often assume that increasing genetic variance always enhances the probabilty of population survival. However, this is not always generally true.' As described here in the case of the natterjack toads genetic variance permits microevolutionary response to environmental change but these responses may not always be adaptive and negative genetic correlations may inhibit evolutionary responses despite a present selection pressure.

6.2 Differentiation in quantitative traits, Q_{ST}

As outlined in Chapter 2, F_{ST} measures the differentiation among populations each subjected to only two selectively neutral processes: the random loss of alleles due to sexual reproduction (i.e. genetic drift) and migration. One way of formulating this is by calculating the proportion of the total genetic variation due to subdivision among populations such as

$$F_{ST} = V_a/(V_a + V_b + V_w)$$

where V_a is the among-sample genetic variance component, V_b is the between-individual within-sample component, and V_w is the within-individual component (Weir and Cockerham 1984).

It is important to note here that F_{ST} only measures neutral differentiation as long as the loci subjected to analyses are truly only evolving through neutral processes. It is technically perfectly possible to calculate F_{ST} also on loci that have been or are subjected to selection. Depending on the nature of selection, the F_{ST} calculated on such loci will deviate from what would be observed at truly neutral loci. Loci subjected to purifying selection due to a common selective pressure would be less differentiated than predicted by neutral loci and loci subjected to local selection unique to each of the subpopulations would be more divergent than predicted by neutral divergence.

For quantitative genetic traits, an analogous statistic of population differentiation has been derived. In such cases population differentiation is calculated as (Wright 1951, Spitze 1993)

$$Q_{ST} = V_{gb}/(V_{gb} + 2V_{gw})$$

where V_{gb} is the additive genetic variance among populations and V_{gw} is the additive genetic variance within populations (Box 6.1).

Now, in analogy to what has been argued for neutral and non-neutral loci, uniform and stabilizing selection over the entire range of the study is implicated if $Q_{ST} < F_{ST}$. Diversifying selection in some or each of the subpopulations is implied if $Q_{ST} > F_{ST}$. When $Q_{ST} = F_{ST}$, the null hypothesis that random processes (i.e. genetic drift) are the cause of the observed divergence cannot be rejected (Fig. 6.3).

6.3 Comparisons of F_{ST} and Q_{ST}

Studies using Q_{ST}–F_{ST} comparisons has increased since the first publication by Spitze in 1993 (Leinonen et al. 2008). Three articles have reviewed studies comparing estimates of Q_{ST} and F_{ST} (Lynch et al. 1999, Merilä and Crnokrak 2001, McKay and Latta 2002). All three found that estimates of Q_{ST} generally exceed F_{ST} in natural populations and thus divergence in quantitative characters seems to be larger than what can be observed using molecular markers (Fig. 6.4). These

Box 6.1 Population divergence in genetic data derived with neutral markers and quantitative characters (after McKay and Latta 2002, Storz 2002)

Assume that that several subdivided populations originate from a common source population that had a genetic variance of $V_{g(0)}$ and that each diverge due to genetic drift. The partititioning of genetic variance at a polygenic quantitative trait within ($V_{g(w)}$) and between ($V_{g(b)}$) populations is related to the partitioning of allelic variation, such as

$$V_{g(b)} = 2F_{ST} V_{g(0)} \quad (1)$$
$$V_{g(w)} = (1 - F_{ST})V_{g(0)} \quad (2)$$

and total genetic variances in the trait

$$V_{g(t)} = (1 + F_{ST})V_{g(0)} \quad (3)$$

and thus

$$V_{g(b)}/V_{g(w)} = 2F_{ST}/(1 + F_{ST}) \quad (4)$$

Storz thus defined a quantitative trait analogue of F_{ST} as

$$Q_{ST} = V_{g(b)}/(V_{g(b)} + 2V_{g(w)}) \quad (5)$$

and therefore $Q_{ST} = F_{ST}$ for a neutral trait.

As pointed out by several authors (e.g. McKay and Latta 2002, Storz 2002), Q_{ST} should be calculated from genetic and not phenotypic variance components since phenotypes are affected by both a genetic and an environmental component. The most common way of obtaining additive genetic variance for a trait is to use common garden experiments and nested ANOVAs with individuals nested within families and within populations.

McKay and Latta (2002) followed Lande (1992) and discussed the evolutionary forces influencing Q_{ST} for a neutral trait. Variation among n populations is determined by migration rates, m, and mutation rates, V_m. Thus

$$V_{g(b)} = ((n-1)V_m)/m \quad (6)$$

However, the variance within populations is driven by the effective population sizes, N_e, and the mutation rate.

Box 6.1 (Continued)

$$V_{g(w)} = 2nN_eV_b \tag{7}$$

Substituting equation 6 into equation 7 yields an estimate of Q_{ST} which can be shown to simplify to

$$Q_{ST} = 1/(1 + 4N_em(n/(n-1))) \tag{8}$$

This is the same as F_{ST} for a large number of demes under an island model. A similar argument can made using coalescence times.

These derivations shows that it is possible to test the null hypothesis that a given trait evolves under neutral processes and that if Q_{ST} deviates from F_{ST} it is possible to reject this null hypothesis.

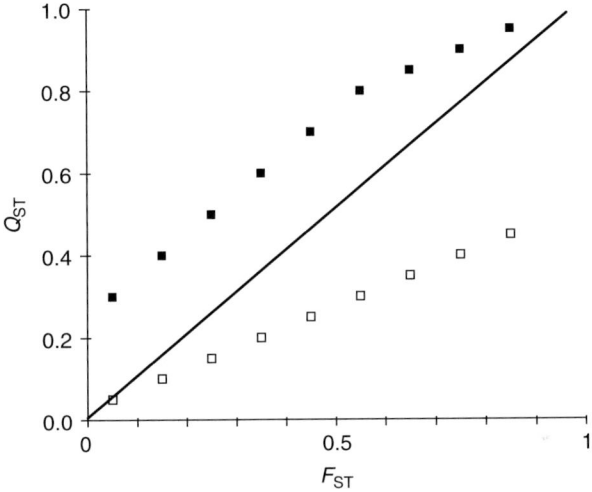

Figure 6.3 Hypothetical Q_{ST}/F_{ST} relationships. White squares describe a situation where $Q_{ST} < F_{ST}$, implying stabilizing selection throughout the range of populations. Black squares describe the case when $Q_{ST} > F_{ST}$ and when diversifying natural selection has caused more differentiation than predicted by random genetic drift (straight line).

results were further confirmed in a meta-analysis by Leinonen et al. (2008): Q_{ST} values are on average higher than F_{ST} analyses (the mean difference being 0.12 units (SD 0.27 units)). The general explanation for this pattern is that natural selection mediated by different local selection pressures has been acting on the genes for quantitative characters, causing a greater divergence.

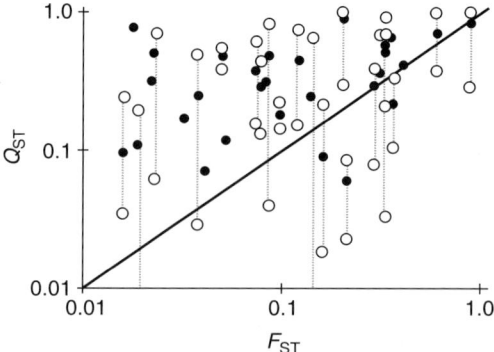

Figure 6.4 Values of F_{ST} and Q_{ST} in 29 species. A filled circle represents the mean value of F_{ST} and Q_{ST} for a given species and the vertical lines and open circles show the large range of Q_{ST} across different quantitaive traits. The diagonal line is the line of equal expectation (from McKay and Latta 2002, reprinted with permission from the publisher).

Populations of grayling, *Thymallus thymallus*, appear to be genetically highly structured over small geographical scales indicating possibilities for local adaptation within the different populations (Koskinen *et al.* 2001). In 1880 a population of grayling was founded in a small mountain lake in Norway and fish from this population were subsequently used for founding populations in two more mountain lakes up until the 1920s (Koskinen *et al.* 2002). Despite small effective population sizes, these populations have been shown to be differentiated in a number of life-history traits including growth rate, survival, and incubation time and perhaps more so than would be predicted from neutral microsatellite markers (Fig. 6.5). These results suggest that life-history traits may rapidly respond to differences in local selection regimes and that if selection is strong enough adaptive differences may evolve in a short period of time. They also suggest that adaptive differences among populations may appear suddenly. If generally true, this is good news for conservation biologists as it would suggest that transplanted populations may quickly evolve towards the optimal phenotype for the local conditions.

However, other studies suggest that a more cautious attitude may be warranted. Studies of two endangered plants, *Brassica insularis* and *Centaurea corymbosa*, showed that Q_{ST} may be smaller than F_{ST} (Petit *et al.* 2001). The authors found high values of θ_{ST} (an F_{ST} analogue) using allozymes in both species, suggesting low amounts of gene flow among the study populations. However, especially for *B. insularis*, Q_{ST} was lower than θ_{ST}, suggesting that the populations studied in each species were experiencing similar selection on the quantitative traits measured. The traits measured were typical morphological traits such as leaf and axis

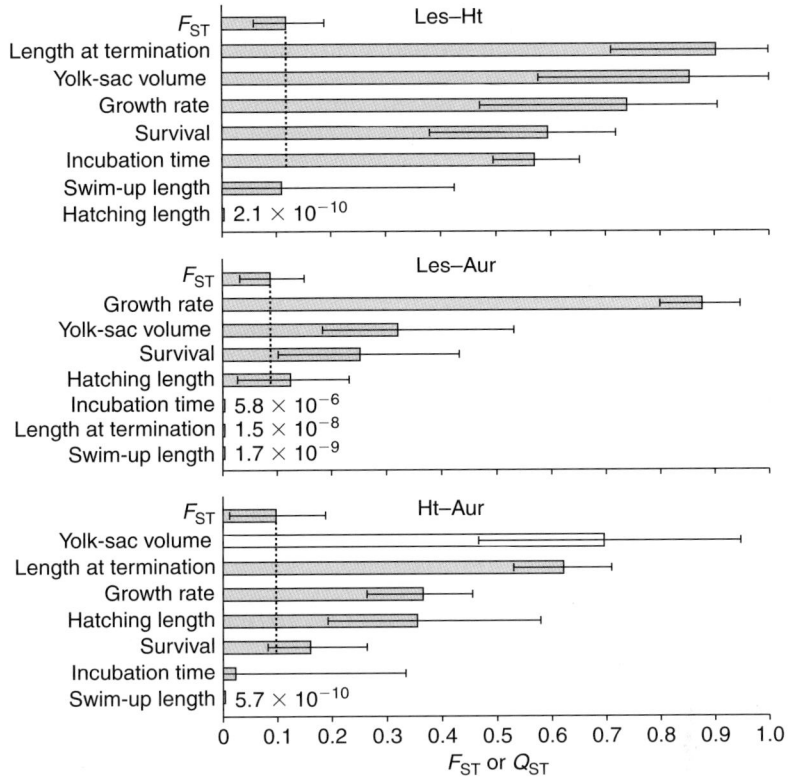

Figure 6.5 F_{ST} and Q_{ST} estimates among populations of Norwegian grayling. Each graph is a comparison of two populations: Lake Lesjaskogsvatn (Les), and Lakes Hårrtjønn (Ht) and Aursjøen (Aur). The horizontal bars indicate 95% confidence intervals (from Koskinen et al. 2001, reprinted with permission from the publisher).

numbers, leaf length and width, and rosette traits but in one case (pubescens in *B. insularis*) a life-history trait reflecting reproductive status was also measured.

The interpretation of any given F_{ST}/Q_{ST} comparison is fraught with problems (O'Hara and Merilä 2005). First, it is important to confirm that Q_{ST} estimates are based on additive genetic variance and not inflated by non-additive genetic effects (dominance and/or epistasis) or environmental effects (Storz 2002). Furthermore, the effects of interactions among loci with different allelic frequencies and dominance relationships remains poorly understood (Goudet and Büchi 2006). Finally there may be biases in the calculation of both F_{ST} and Q_{ST}. Epistatic variance may inflate estimates of Q_{ST} (Lynch et al. 1999) and the upper bound of F_{ST} becomes deflated when many variable loci are used in calculations of F_{ST} (Hedrick 1999). It has further been pointed out that as variation for quantitative traits is introduced by mutation at a higher rate than for molecular markers, the extent of variation for quantitative traits usually exceeds that for molecular markers (Lynch *et al*. 1999). These authors further stressed that population divergence due to stochastic processes such as drift and founder effects may be detectable with quantitative traits but not with molecular markers. This is because some molecular markers (in particular allozymes) may show low levels of polymorphism and may thus be fixed among populations. There are also statistical problems related to the bias and precision of estimates of Q_{ST} (and F_{ST}; O'Hara and Merilä 2005).

Among these problems it appears as though dominance is a minor problem (Goudet and Büchi 2006). Many of the others are still unresolved. However, the statistical issues may be solved by sampling more populations than usual in empirical studies (i.e. many more than seven) and by employing proper resampling techniques in calculations of estimates and their variances (O'Hara and Merilä 2005).

6.4 Q_{ST} applied to conservation studies

Studies of the effects of human-induced habitat fragmentation on both neutral and adaptive genetic variability are still scarce. In common garden experiments using common frogs, *R. temporaria*, from populations from continuous or fragmented parts of the species distribution in southern Sweden, positive relationships between mean values of fitness-related traits (survival probability of eggs and froglets and body size) and the amount of microsatellite variation in a given population were found (Johansson *et al*. 2007). F_{ST} tended to be more pronounced in the fragmented than in the continuous habitat, indicating that habitat fragmentation increases neutral population structure as expected. However, in the continuous habitat but not in the fragmented habitat Q_{ST} exceeded F_{ST} (Fig. 6.6). These results suggest that the impact of random genetic drift relative to natural selection was higher in the fragmented landscape where populations were small, and had lower genetic diversity and fitness compared with populations in the more continuous landscape.

However, the outcome of F_{ST}/Q_{ST} comparisons may be unpredictable. For both of two endangered plant species, *B. insularis* and *C. corymbosa*, Q_{ST} values were smaller than F_{ST} and in each case Q_{ST} was independent of geographical distance (Petit *et al*. 2001). This was in contrast to positive isolation by distance in F_{ST}. This suggests that using F_{ST} as a proxy for the lower bound for Q_{ST} may overestimate the evolutionary potential of these endangered species. The authors suggest that for endemic species with restricted distributions, the ecological niche is often restricted and homogeneous selective forces are likely to act on the populations.

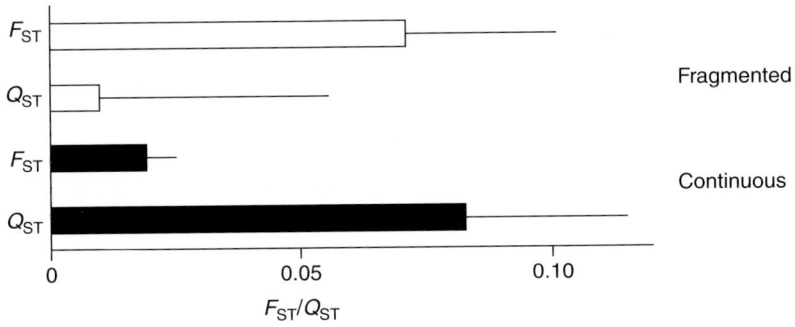

Figure 6.6 Estimates of among-population differentiation (±1 SE) in body size (dry mass, Q_{ST}) and seven microsatellite loci (F_{ST}) in fragmented (open bars) and continuous landscapes (filled bars) (from Johansson et al. 2007, reprinted with permission from the publisher).

On the other hand, small population sizes and restricted dispersal may in these cases produce strong differentiation for neutral variation. It appears that in this case reinforcement or reintroduction programmes would be good strategies for restoring these endangered species. However, the authors are cautious to point out that outbreeding depression may still be a result of such efforts if other unstudied locally coevolved gene complexes exist.

As mentioned above, to employ a proper estimation of Q_{ST} it is required that the variance of the quantitative trait is measured as the additive genetic component for the trait, otherwise the Q_{ST} estimate may be inflated by environmental factors (Storz 2002). The most common way to estimate additive genetic variance is the conduct common garden experiments and to estimate heritabilities among half-sibs. However, for many endangered and threatened species this is not feasible or simply not possible. Some authors have therefore used the phenotypic variances in which the variance components are obtained from a standard analysis of variance with population as the factor.

Storz (2002) used this approach in a study of body-size variation among populations of Indian fruit bats, *Cynopterus sphinx*. To obtain meaningful comparison he assumed the heritability of body size to be 0.5 (Falconer and MacKay 1996). It was found that to explain the observed differences among the populations with genetic drift, excessively high levels of heritability had to be assumed.

We applied a similar phenotypic $Q_{ST}(P_{ST})/F_{ST}$ approach to study spatial variation in selection among great snipe, *Gallinago media*, populations in two regions in northern Europe (Saether et al. 2007). Morphological divergence between regions was high despite low differentiation in selectively neutral genetic markers. However, populations within regions showed very little neutral divergence and trait differentiation. To be able to conclude that $Q_{ST} > F_{ST}$ we tested the robustness of the result by sensitivity analyses in which we varied the assumptions about additive genetic variance underlying the traits. We found that our conclusion was indeed robust against altering assumptions about the additive genetic proportions of variance components. The homogenizing effect of gene flow (or a short time available for neutral divergence) has apparently been effectively counterbalanced by differential natural selection in two traits: tarsus length and tail white (Fig. 6.7). Bill length showed some evidence of being under uniform stabilizing selection among the populations but testing whether Q_{ST} was indeed less than F_{ST} in this case was difficult because the F_{ST} was close to zero and because of the variance in Q_{ST}.

The implication of this and the other studies reviewed above is that neutral markers can be misleading for identifying evolutionary significant units but that the Q_{ST}/F_{ST} approach has many pitfalls and that the conclusion depends on the nature of the trait. It is important to have some *a priori* understanding of the biology and the nature of the traits and the selective regimes before any firm generalization about Q_{ST}/F_{ST} comparisons are drawn in a conservation context. Traits such as those related to photoperiodism may be predicted to vary with latitude but populations residing on the same latitude may respond similarly even when other factors differ among populations. On the other hand, if some factor covaries with the adaptive response to photoperiodism, such as altitude, it may be that local selection creates a different response to day length in such populations.

Many species of conservation concern would not lend themselves to common garden experiments even if such is the most stringent way to conduct a Q_{ST}/F_{ST} study. With proper sensitivity analyses a P_{ST}/F_{ST} approach might be valuable when common garden experiments are not an option.

6.5 Conclusions

Local adaptation is common in nature and therefore probably ubiquitous in threatened populations subjected to conservation efforts. Local adaptations can be detected by either the direct allophonic method or by the indirect synchronic method. Another way of detecting and studying local adaptation is to compare the divergence in quantitative traits Q_{ST} with that obtained from the F_{ST} values of neutral markers. If $Q_{ST} > F_{ST}$ this might be taken as indicative of natural selection causing greater divergence among populations than the background imposed by genetic drift. A goal in conservation studies should be to reach and manage natural populations so that they can respond to local selection and adapt to the environment in which they live. In natural populations effective size (smaller

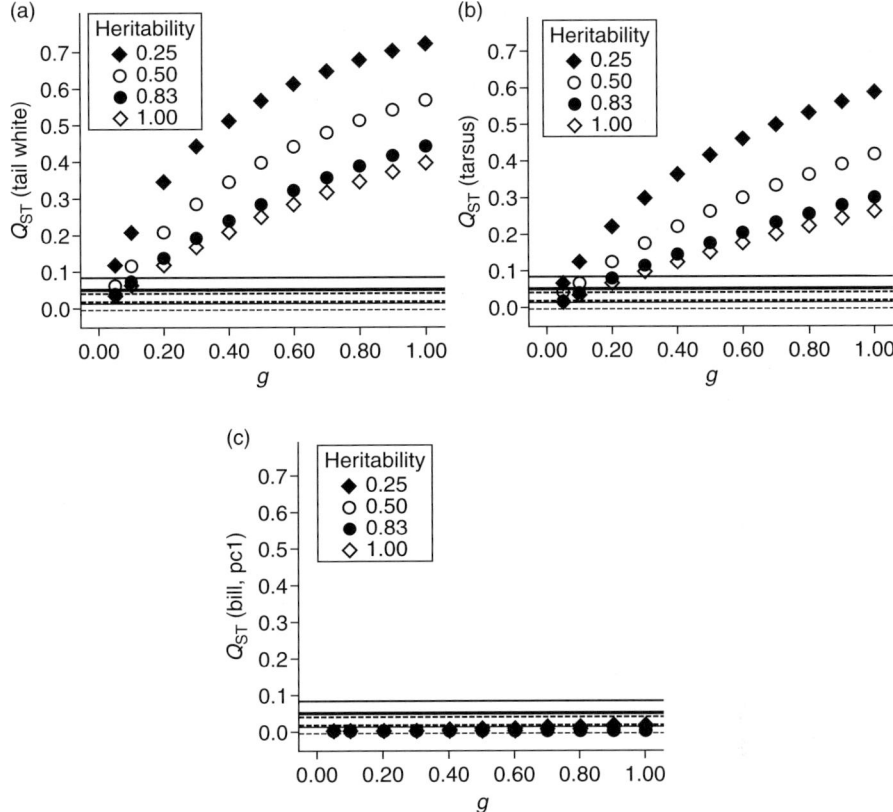

Figure 6.7 Q_{ST} sensitivity plots of varying the additive genetic proportion of between-population (g) and within-population (heritability) variance components. Q_{ST} values are calculated treating two regional distributions of great snipe as two populations. Estimates of neutral divergence are shown as horizontal lines with 95% bootstrap confidence limits (bold solid line for R_{ST}; dashed line for Weir–Cockerham F_{ST}). Q_{ST} for (a) tail white and (b) tarsus is considerably larger than neutral genetic divergence for most parameter space, whereas Q_{ST} for (c) bill (pc1) is similar or smaller (from Saether et al. 2007, reprinted with permission from the publisher).

size means the less variation), mutation, and selection interact to determine the amount of variation maintained (Lande and Barrowclough 1987). As discussed in Chapter 1 the recommendation is that an N_e of more than 500–5000 should be targeted as a practical goal (Franklin 1980, Lande and Shannon 1996).

This and the previous chapter have revealed a controversial issue and apparent conflict in conservation biology: should conservationists be aiming at preserving and restoring genetic variation as such or should the focus be on preserving local adaptations? This depends on the nature of the variation and the type of selection imposed on a population (Lande and Shannon 1996). Imagine a melanic population of *Peromyscus* mice living on a dark substrate threatened by local extinction. Clearly conservationists would hesitate to introduce light-pelaged mice in to this population. Even if genetic variation would be boosted, so would also the maladapted pelage genes. This example is illustrative because the link between coat colour and the environment is obvious and clearly visible. However, natural populations are also likely to be adapted to local selection regimes in more subtle and less clearly understood ways. Therefore conservationists should be careful when transplanting organisms, it should only be considered when the target population clearly suffers from low effective population size and shows signs of inbreeding depression.

7 Ecological genomics

The first decade of the twenty-first century has been called the age of 'omics'. The now famous word ending was first used in genomics but now transcriptomics, proteomics, and the like are also used. The first whole genome sequenced was that of *Haemophilus influenzae*, the genome sequence of which was published in 1995 (Fleischman *et al.* 1995). By the end of 2007 more than 700 completely sequenced genomes were available (see GOLDTM, the Genomes Online Database, v 2.0 at www.genomesonline.org) and more than 3000 whole-genome sequencing (WGS) projects are on the way. Most of the published genomes are bacterial. In eukaryotes, WGS projects are focusing primarily on fungi, protists, and plants but other taxa are also being subjected to WGS. Sequencing of more than 40 mammalian species and six bird species is now in progress (Segelbacher and Höglund 2008). WGS projects focus primarily on so-called model organisms for genetic and physiological research and on species of economical or agricultural interest. However, some whole-genome projects have been chosen because the species have a phylogenetic position that makes the data gathered useful in comparative genomic projects.

It is easy to get carried away by the technical advances and the landmark findings that are reported on a regular basis in the weekly science journals. No doubt, the availability of whole-genome information opens new research fields which can be of interest not only in model species, but can also be of potential use in related species. As more and more species become sequenced, the comparison of genomes will enable the identification functional DNA regions in ecologically interesting species (Travers *et al.* 2007, Piertney and Webster 2008, Wheat 2008). Whether these kinds of data and the techniques they allow to be employed will ever be of much use in the study of endangered or rare species is less certain. Ecological and evolutionary applications of genomics are still in their infancies and it is far too early to make any good guesses about future research directions. At the moment, the financial costs for a WGS exceed what is usually spent on an average endangered species and I doubt that any but a few conservation flagship species will be subjected to WGS. However, as I hope will become apparent in this chapter, some of the tools and some of the techniques may also be applied to species about which there is or will be only limited genomic information.

7.1 WGS

Genomics is the science of whole genomes. The focus of interest is on the properties of entire and completely sequenced genomes, such as genomic architecture, size of genomes, number of genes, gene order, and synteny (Pagel and Pomiankowski 2007). Typical questions are: how is the genetic information compartmentalized? What is the extent of regulatory genes? What is the extent of informative versus junk DNA? Are there any transposomal elements present? Although interesting and perfectly valid research foci, it is clear that these are quite far from the concerns of the average conservation biologist. However, given genomic information, a number of facts useful to conservation may be extracted. I will review these applications later on in this chapter but first I will briefly describe the techniques for gathering the data and types of analyses involved to get to this first step.

The first thing to do in any genomic survey is, of course, to gather masses of sequence data. In model organisms, the template for such studies are often a few isolates or an inbred line of the species of interest. Until recently, WGS projects used traditional Sanger sequencing which, depending on the organism, involved massive cloning and sequencing of mega-base pairs. Even for modest-sized genomes, these projects took years to complete and involved huge consortia. Currently, there are tremendous advances in fast and inexpensive sequencing technologies (Wheat 2008). Recently it has been established that alternatives to Sanger sequencing, such as parallel pyrosequencing, can provide deep coverage of eukaryotic genomes and transcriptomes (Bainbridge *et al.* 2006, Cheung *et al.* 2006, Weber *et al.* 2007, Vera *et al.* 2008). These major technological developments make it feasible to collect genomic data from non-model species.

Whereas it may not be feasible to invest heavily in genome sequence resources for every species or natural population of interest in conservation biology, so-called expressed sequence tags (ESTs) are a relatively inexpensive genomic resource that can be developed for almost any organism regardless of sequencing strategy (Bouck and Vision 2007). ESTs are single-read sequences produced from partial sequencing of an mRNA pool. Reverse transcriptase is used to produce cDNA, which is then cloned into a vector library and sequenced. EST libraries thus provide a snapshot of the transcribed mRNA population within a given set of tissues, developmental stages, environmental conditions, and genotypes (Rudd 2003, Dong *et al.* 2005). Previously, transcriptome data was obtained by Sanger sequencing of ESTs (Adams *et al.* 1991). Despite improvements to Sanger sequencing over the past 30 years, this methodology is still labour-intensive and expensive. By contrast, a single 8-hour sequencing run using 454 pyrosequencing (or similar techniques) can generate mega-bases of DNA sequence and does not involve any cloning step (Margulies et al. 2005). Parallel pyrosequencing yields randomly fragmented sequencing reads that, if sufficiently abundant, may span the entire transcriptome (if based on mRNA) or genome (if based on DNA).

The average conservation biology laboratory is probably equipped to perform Sanger sequencing in house or to prepare samples so that they can be sequenced commercially at reasonable cost. However, at the time of writing, it is necessary for most conservation biologists who are contemplating parallel pyrosequencing to team up with a specialist laboratory that has the equipment and knowledge to perform such studies.

7.2 What to do with the data? Assembly and annotation

Regardless of method, genomic studies produce masses of sequence data and the assembly and annotation of such data are not trivial tasks. As an example, *de novo* assembly of a eukaryote transcriptome using 454 pyrosequencing data has established the utility of gathering such data in an ecologically well-studied but genomically unknown species, the Glanville fritallary, *Melitaea cinxia* (Vera et al. 2008), as outlined below.

There are a number of specialized statistical tools and software packages available for this assembly stage (see Wheat 2008 for a review). In short, raw sequences are filtered so that low-quality reads are taken out of the data; the remaining ones are entered into the assembly. Next the high-quality reads are aligned and overlapping sequences are combined into contiguous sequences (so-called contigs). Non-overlapping reads are left as singletons.

The Glanville fritallary study used two normalized complementary DNA collections from about 80 individuals collected in the study area, including larvae, pupae, and adults. Using 454 sequencing they produced 608 053 ESTs of which 518 079 exceeded the minimal quality-standard filtering and entered the assembly. The ESTs were of a mean length of 110 nucleotides. This assembled into 48 354 sets of overlapping DNA segments (contigs) and 59 943 single reads. For quality-control purposes they also used Sanger sequencing to obtain 3888 sequence reads from Glanville fritillary cDNA libraries. With this technique they found 364 contigs with an average length of 574 bp. In general they found good agreement between 454 ESTs and ESTs obtained by traditional methods.

The authors then compared their data with sequences already banked in Internet databases from other organisms like *Drosophila* species, the genomically well-studied silkworm *Bombyx mori*, and the butterfly *Heliconis erato* and confirmed the accuracy of the sequencing and assembly. These comparisons allowed the authors to find about 9000 unique genes and more than 6000 additional unannotated contigs. These unannotated contigs were confirmed to be expressed genes by microarray analyses. The average depth of the coverage was 6.5-fold, meaning that any sequence in the transcriptome was sequenced about six times. This example shows the utility of genetically well-studied genomic reference species for the annotation part of the process.

7.3 What to do with the data? Evolutionary and ecological analyses

When there is an annotated assembly the next issue becomes: what to do with it? As noted above simple descriptive statistics of a genome or a transcriptome do not aid a conservation project very much. Fortunately, whole-genome or transcriptome sequencing enables functional genomic studies. Such studies have so far by necessity been applied mainly to a few model organisms. However, the conservatism in gene organization and in the sequences of functional genes give hope that such tools may also be used in non-model species. Some, so-called housekeeping genes, code for gene products that are involved in cell-physiological processes that have been retained and maintained in many life forms for millions of years. It is already obvious that the findings in model organism research may be applied to closely related species; for example, findings in *Arabidopsis thaliana* can be applied to and tested in other species in the genus. Similarly, findings in studies of the domestic chicken, *Gallus gallus*, can be applied to other galliforms. Recently quantitative trait loci (QTL) for female comb size were shown to be non-randomly associated with female reproductive investment in domestic chicken (Wright *et al.* 2008). Whether such results can be applied to non-models and how phylogenetically distant they can be from a given model organism depends on the conservation of the particular sequences under study.

At present there are very few large-scale genome studies on ecologically relevant non-model organisms in which questions about adaptive genetic diversity in natural populations can be addressed. This problem may be exemplified by birds. Birds are very well known ecologically since there has been a long research tradition of ecological studies. In birds, the species with the most information on genomic architecture and gene sequence are the galliform domestic chicken (www.ncbi.nlm.nih.gov/projects/genome/guide/chicken/) and the passerine zebra finch (http://songbirdgenome.org/). However, these model systems may be poor indicators of genomic resources in other species as chickens have gone through multiple generations of domestication, and passerines are evolutionarily quite distinct from other bird taxa.

Under the present biodiversity crisis conservation biologists need tools to define taxa and prioritise populations that need protection in the face of limited resources. It has been claimed that these so-called management or evolutionarily significant units need to be defined as populations and lineages that are demographically and hence evolutionarily independent. How best to define such units is unclear (Crandall *et al.* 2000). Beaumont and Balding (2004) highlighted that 'Hitherto, the degree of adaptive divergence between populations has been determined genetically by some measure of distinctiveness—for example the possession of reciprocal monophyly in mitochondrial sequences. However, this distinctiveness may, particularly if only based on [mitochondrial] DNA or a few nuclear markers, largely reflect the vagaries of demographic history. What is needed is to be able to quantify the distinctiveness of populations in terms of their local adaptation..., which may also only involve a few genes, but genes with key functional roles' (see also Luikart *et al.* 2003).

The issue then boils down to the identification and localization of the genes underlying fitness differences and adaptive divergence in natural populations (Luikart *et al.* 2003, Vasemägi and Primmer 2005, Butlin 2008, Piertney and Webster 2008). One way to do this would be to test whether it is possible to find genetic variation that correlates with fitness in natural populations in genes that have a known function in a genomic reference species (i.e. a candidate gene approach). However, with this approach it is impossible to find new genes of functional importance in non-models. At the other extreme are studies using genome scans to detect signs of selection (e.g. Cork and Purugganan 2005; Fig. 7.1). Purifying and diversifying selection on polygenic traits can be expected to produce predictable patterns of allelic variation at the underlying loci underlying a QTL, and the locus-specific effects of selection should therefore be detectable against stochastic variability of the rest of the genome (Storz 2005). Vasemägi and Primmer (2005) called for the use of a set of complementary research strategies to find functionally important loci (Box 7.1).

It is clear that there no single strategy to cover the whole way of understanding the genetic basis of ecologically important traits. Basically there are two rather fundamentally divergent research traditions that need to be merged. On the one hand researchers in ecological genetics have long been using the quantitative genetic tools of plant and animal breeders. As such, additive genetic variation in the form of heritabilities has been established for many important life-history traits in wild populations (see Chapter 2). New statistical advances in quantitative genetics have revitalized the study of quantitative genetics in natural populations (Frentiu *et al.* 2008, Ovaskainen *et al.* 2008). However, when it comes to identifying the genomic regions and the genes underlying the variation in these life-history traits, less progress has been made (but see examples below). In going from genes to ecology, the research traditions of molecular genetics, functional

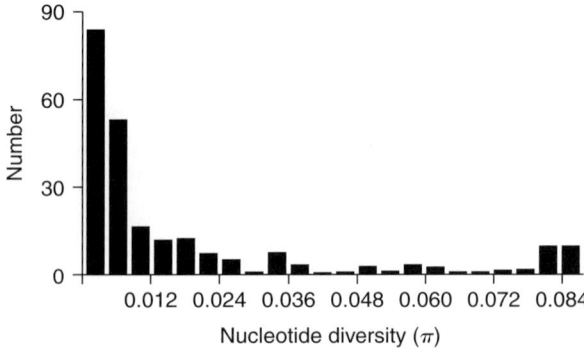

Figure 7.1 Nucleotide diversity for dimorphic genes in *Arabidopsis* (from Cork and Purugganan 2005, reprinted with permission from the publisher).

Box 7.1 Methods to detect functionally important genetic variation

---

The methods have been categorized on a scale from bottom-up to top-down approaches depending on the focus of the research along the genotype–phenotype pathway (after Vasemägi and Primmer 2005).

Single-locus and sequence-based 'neutrality' tests

Listed below are statistical tests designed to test whether a particular DNA sequence have evolved under a neutral model or under stabilizing or balancing selection. Example of tests used are listed below.

- Deviations from the expected Hardy–Weinberg genotypic proportions within a population (Watt and Dean 2000).

- The Ewens–Watterson test uses the allele frequency distributions and tests if there is more linkage disequilibrium and less genetic variation in the particular region than is expected in a neutral marker.

- Tests to detect evidence of selection in the past such as the Hudson–Kreitman–Aguadé (HKA) test and Tajima's D test are based on the distribution of sequenced alleles and/or the level of sequence variability (Watterson 1977, Hudson *et al.* 1987, Tajima 1989, Fu and Li 1993, Fu 1996, and Fay and Wu 2000).

- Tests based on the non-synonymous and synonymous substitution ratio (dN/dS or KA/KS) and McDonald–Kreitman-type test (Hughes and Nei 1988, McDonald and Kreitman 1991). The McDonald–Kreitman test is

Box 7.1 (*Continued*)

based on the observation that if the observed variation is neutral, then the rate of substitution between species and the amount of variation within species are both a function of the mutation rate. Thus the ratio of non-synonymous to synonymous fixed differences between species should be the same as the ratio of non-synonymous to synonymous polymorphisms within species

Reviews of statistical tests that can be used to test for selection on DNA sequences are found in Kreitman and Akashi (1995), Kreitman (2000), Otto (2000), Ford (2002), and Garrigan and Hedrick (2003).

Tests of dN/dS ratios are reviewed in Nielsen (1997), Yang and Nielsen (2000), and Bierne and Eyre-Walker (2003). Some of these tests may explicitly examine which amino acid sites that have been subjected to selection (see examples in Nielsen and Yang 1998, Yang *et al.* 2000, Suzuki 2004, and Massingham and Goldman 2005).

Multiple-marker-based 'neutrality' tests

Information from many loci may be used to test whether any loci deviate from a neutral null distribution of variation. The studies of amplified fragment length polymorphisms (AFLPs) and environmental variation in periwinkles and common frogs are examples of this (Wilding *et al.* 2001, Bonin *et al.* 2006).

The test of Lewontin and Krakauer (1973) examines the variation among populations for given loci. Theoretically, all loci are subjected to the same amount of genetic drift and gene flow. Thus the expected variance over populations should be the same. However, differential selection among populations increases the variance. On the other hand, purifying selection over populations decreases the variance. This test have been used to identify outlier loci (see Luikart *et al.* 2003, Storz 2005).

QTL mapping of mRNA expression variation

Linkage mapping of mRNA transcripts are used to identify particular regions of the genome that are associated with variation in gene expression levels (Jansen and Nap 2001, Doerge 2002). This method requires that microarrays are developed for the study species or a reasonably closely related species and thus this method may be of limited value for conservation biologists.

Box 7.1 (*Continued*)

Allele-specific mRNA expression analysis

Estimation of expression levels of alternative alleles within heterozygous individuals based on polymorphism in the transcribed region of a gene (Buckland 2004, Knight 2004, Yan and Zhou 2004).

QTL mapping of protein expression variation

Linkage mapping and protein expression analysis are used to identify particular regions of the genome that are associated with variation in quantitative and qualitative protein expression levels within a pedigree (Gorg *et al.* 2004). This approach, as with the previous one, suffers limitations such as a requirement of a large amount of pedigree material and a reasonable amount of fresh tissue and may thus be of limited value in conservation studies. However, the study of gene regulation has proven important in studies of various stressors such as drought (de Vienne *et al.* 2001), which is of direct relevance for conservation.

Environmental association analysis

This analysis estimates significant associations between environmental variables and specific alleles. Such may be taken as evidence for directional selection affecting a particular locus. Studies can be temporal (by following cohorts in time) or spatial, even at small spatial scales (Johannesson *et al.* 1995).

QTL analyses (linkage mapping)

If there is genetic linkage map information (i.e. knowledge where and on which chromosomes markers are positioned), pedigree material to trace the segregation of the markers, and phenotypic data of individuals in a pedigree, it is possible to tests for association between markers and certain ecological traits of interest (Erickson *et al.* 2004, Slate 2005). This technique has been extensively used in model organisms and domesticated species (Andersson and Georges 2004).

Admixture mapping

This is a similar technique to the one above but here the association among and between populations and their experimental backcrosses are used to identify traits and genomic regions that are distributed non-randomly (Rieseberg and Buerkle 2002, McKeigue 2005).

> Box 7.1 (*Continued*)
>
> **Association analysis (linkage disequilibrium mapping)**
> Tests of a non-random association of a phenotypic trait of interest within families or populations and a certain genotype (or haplotype) by utilizing the non-random occurrence of alleles at linked loci, known as linkage disequilibrium (LD). This approach requires that a large part of the genome is covered by a large set of genetic markers. For example, in humans it has been suggested that 1 million random single nucleotide polymorphisms (SNPs) are needed to provide reasonable whole-genome coverage for association studies (Hirschhorn and Daly 2005). Clearly this is at present unreasonable for most species of conservation concern. However, most studies of non-model species using this approach focus on LD between a limited number of candidate genes. For example, associations between major histocompatibility complex (Mhc) genes and immune response have been identified (reviewed in Bernatchez and Landry 2003 and Garrigan and Hedrick 2003).

genetics, and genomics need to be incorporated. In this research tradition there has been a focus on a few model organisms that are often poorly characterized from an ecological standpoint. It goes without saying that none of the model organisms belong to the category of threatened species about which conservation biologists are concerned.

Luikart *et al.* (2003) introduced the concept of population genomics, which was defined as 'the simultaneous study of numerous loci or genome regions to better understand the roles of evolutionary processes (such as mutation, random genetic drift, gene flow and natural selection) that influence variation across genomes and populations.' They proposed the following research strategy: step 1, sample as many individuals from as many populations as possible; step 2, genotype as many loci as possible, preferably with an even spread throughout the genome; step 3, test for outlier loci (such as loci that have a greater- or smaller-than-average F_{ST} values); step 4, on the neutral loci, compute evolutionary or demographic parameters without using outlier loci. On the candidate (adaptive) loci, test for causes of outlier behaviour (for example, selection) and use adaptive information for biodiversity conservation or evolutionary inferences (Fig. 7.2).

As an example of a study using the research outline by Luikart *et al.* (2003), studies from my own research group of willow grouse, *Lagopus lagopus*, may

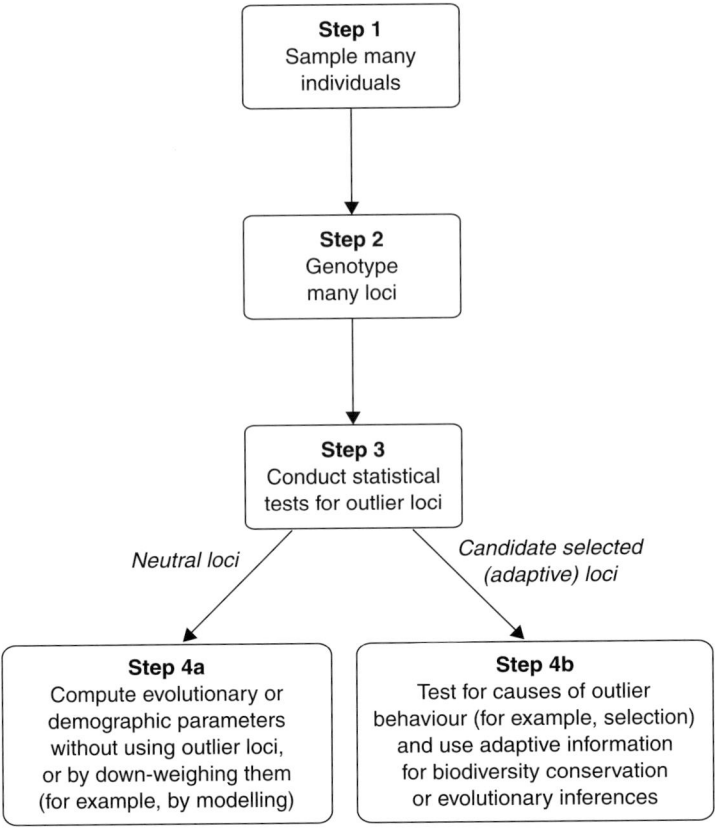

Figure 7.2 Flow chart indicating the steps in a population genomic investigation (from Luikart *et al.* 2003, reprinted with permission from the publisher).

serve as an example. We sequenced 18 autosomal protein-coding loci from approximately 15–18 individuals in four populations (S. Berlin, M. Quintela, and J. Höglund, unpublished results). From these sequences we retrieved more than 100 independently segregating single nucleotide polymorphisms (SNPs; see below). We found unusually high levels of nucleotide diversity in Scandinavian willow grouse as well as very little population structure among localities that were up to 1647 km apart. None of the loci diverged from neutral expectations. There were also low levels of linkage disequilibrium, even within the genes, and the population recombination rate was high, indicative of an old panmictic population, where recombination has had time to break up large haplotype blocks. Compared with the silent nucleotide diversity at third codon position, the non-synonymous nucleotide diversity was low, which is in agreement with effective purifying selection, possibly due to the large effective population size. In birds nucleotide-level variation is poorly characterized; the domesticated chicken and a few passerine species are exceptions in this respect (Backström *et al.* 2006). However, these studies suggest that bird nucleotide diversity is high (approximately 10^{-3}) and a likely explanation for this is the generally higher effective population sizes compared with mammals. Our findings in the willow grouse need to be repeated in other bird species and such studies should increase the number of both synonymous and non-synonymous SNPs. At present the non-synonymous substitutions are far too few to address any relevant questions about patterns of adaptive genetic variation in willow grouse or any other bird species. With more markers and more species studied it will be possible to make general conclusions of levels of genetic variation in natural bird populations and to test the hypotheses about the distribution of adaptive and neutral variation. If local adaptation is important in birds it is predicted that local populations and subspecies will show higher levels of differentiation in adaptive genes than in neutral. If this is the case, threatened bird populations should be managed accordingly.

7.4 Genomics in conservation

Genomic applications and techniques are likely to become more and more prevalent as genomic data from endangered and related species become available. As noted previously in this chapter it is not likely that data from endangered species will lie at the forefront of ecological genetics but experimental conservation genetic studies are often conducted on genetically well-known species such as *Drosophila* (e.g. Bijlsma *et al.* 1999, 2000). In these circumstances genomic resources and techniques may turn out to be useful.

7.4.1 SNP detection and genotyping

SNPs have a number of features that make them desirable genetic markers in studies of genetic variation in natural populations. In a review of the use of SNPs in conservation studies, Morin and coworkers (2004) listed a number of applications number of applications where SNPs are likely to become useful. While SNPs are less polymorphic than the major alternative microsatellites, they are more abundant throughout the genomes. Furthermore SNPs evolve by a simpler mutational process as compared to microsatellites. The short stretches of repetitive DNA sequences of microsatellites evolve when the endogenous DNA polymerase in the cell makes a replication error and either misincorporates or mistakenly removes a copy (so-called DNA slippage). These stepwise mutations are much more likely than many other types of mutation (in the order of 10^{-3}

instead of 10^{-6} per genome and generation). The mutation process of SNPs, on the other hand, is simpler, as they evolve mainly by point mutation. By being less variable, more SNPs are needed than microsatellites to achieve resolution power in many applications. However, by being less variable the problems of homoplasy (the same allelic state evolving more than once in the sample) can effectively be avoided.

Protocols for developing microsatellite markers are quite straightforward and affordable for any conservation genetic study anticipating studying genetic variation. However, microsatellites seem to be rare and hard to develop in genomes of some organisms (e.g. butterflies and arthropods). Briefly, microsatellites are found by cutting up the genomic DNA from the study species with restriction enzymes and ligating this DNA into bacteria using a phagemid vector. This genomic library is then probed with a synthetic oligonucleotide mirroring a repeat sequence. The bacterial clones are allowed to grow and clones with positive inserts are sequenced. This will allow detection of not only the repeat sequence but also flanking regions around the repeat. In the next step primers in the flanking regions close to the repeat sequence are designed. With the aid of the primers, the target DNA can be manifolded in a PCR to screen allelic length variation in a large number of individuals.

As indicated above SNP discovery is a more complicated process and several strategies exist (see Morin *et al.* 2004, Slate *et al.* 2008 for reviews). One method is referred to as exon priming intron crossing (EPIC). Even in the absence of previous sequence data, primers can be designed by aligning of sequences in exons of protein-coding genes in related species that have been sequenced. A pair of primers are made of which the forward primer is designed in one exon and the backward primer in an adjacent exon so that the intron in between can be amplified and sequenced. By sequencing intronic DNA the chances of discovering segregating SNPs are maximized. A related approach is to use core anchor tagged sequences (CATS). This is a set of primers originally developed for gene mapping in diverse set of organisms and they were therefore chosen to be as conservative as possible. Some but not all of these primer pairs yield PCR products, mostly within coding genes (Lyons *et al.* 1997).

The second main approach to find SNPs is to sequence random genomic fragments in a limited number of individuals and the aligning the sequences. SNPs are found by identifying segregating sites in the alignment. Random clones from genomic DNA libraries may be sequenced or existing EST databases may be mined for SNPs using a number of different computer programs. Obviously 454 sequencing will be useful in this endeavour.

SNP genotyping can be performed in house, but in a recent review outsourcing to specialized laboratories was recommended (Slate *et al.* 2008). A number of strategies and platforms for SNP genotyping are available and the choice of these depends on the number of samples and SNPs. High-throughput genotyping is made possible by the fact that most SNPs are biallelic, which can be utilized to streamline the genotyping to accommodate a large number of samples and many SNPs fast and cost-effectively.

SNPs can be used for estimating genetic variation. It is believed that by using a large number of SNPs a better and more representative estimate of genomic diversity may be obtained. However, this increased precision comes at a cost. Reliable estimates of genome-wide variation required four to ten times more biallelic amplified fragment length polymorphism (AFLP) markers with multi-allelic markers (Mariette *et al.* 2002). However, dominant AFLP markers are less informative than codominant biallelic SNP markers and thus fewer SNPs may be required compared with AFLP loci.

SNPs can be used in identifying individuals and to reveal parentage and relatedness. There are already established techniques for this using microsatellites and AFLPs, and it is not clear whether these applications would be improved by using SNPs. On the other hand, there are no indications that such analyses would be worsened by using SNPs.

Similarly, SNPs may be used in estimates of population structure. Estimating F_{ST} from microsatellites can be problematic. Hedrick (2005) showed that the theoretical upper limit of an F_{ST} estimate is not 1 (as established for biallelic loci); instead, the upper limit of a multilocus estimate (G_{ST}) may be considerably lower and Hedrick suggested a correction to remedy this effect. Since most SNPs chosen for analysis are biallelic this problem may be less significant for these markers.

Estimating changes in past population size as reviewed in Chapter 4 is one area in which SNPs will not be an improvement. This is because these tests are more powerful with a higher the number of alleles at a locus.

7.4.2 QTL mapping of functionally important loci

As reviewed in Chapter 2, quantitative trait variation always has a genetic component (large or small depending on the trait and circumstances). It has been argued that traits subject to natural selection, and thus representing adaptations to local conditions (Chapter 5), are of particular importance in conservation. Finding and characterizing the molecular basis of QTLs is therefore potentially important.

The application where SNPs are used to their best advantage is in mapping of QTLs, because SNPs can be typed on a much larger scale and are much more abundant than microsatellites. Using SNPs, any genomic location may thus be analysed (Slate *et al.* 2008). Mapping means going from QTL location to finding candidate genes and is seemingly a simple process. Mapping experiments in model organisms such as laboratory rats, tomato, and cattle have shown that complex traits in inbred organisms may have a simple genetic architecture, with only a handful of chromosomal regions that associate with any given trait (Flint and Mott 2001). However, we are a long way from providing similar results in non-model species. Even in model species it is difficult to find an statistical association between a marker locus and a trait. Markers typically are linked to an approximately 30 cM region of a given chromosome and thus even if a candidate gene is found within that region there may be several candidates and so finding the responsible mutation, the quantitative trait nucleotide, is hard and requires complex research protocols and large sample sizes. At the very least, to map complex traits in natural populations a linkage map is required so that the chromosomal position of the markers is known. Linkage maps have started to appear in some non-models (Hansson *et al.* 2005) but are likely to be rare in the average threatened species subjected to a conservation genetic study.

7.4.3 Differential gene expression

Microarrays are generally used to quantify differences in global gene-expression patterns between groups of individuals. In conservation genetics, microarrays can be used to screen the transcriptome for genes that might be differentially expressed in relation to specific treatments or coming from different populations (e.g. a threatened and a viable one). Therefore, this is a tool for detecting candidate genes for further study (e.g. Whitehead and Crawford 2006).

A classical microarray is a sample of small dots of known DNA (whole genes or parts of genes), usually collected on a glass plate or a silicon chip, to which cDNA or RNA can hybridized in a dot specific manner. The hybridizing material is labelled with fluorescent dyes and hybridized to the DNA dots. If the DNA sequence on the chip has been expressed, a corresponding sequence is present in the test sample and the relative expression levels can be read as relative intensities of the dyes for each dot. Several statistical tests have been utilized to identify differentially expressed genes from two different EST libraries (reviewed by Ruijter *et al.* 2002).

Kristensen *et al.* (2005) used a microarray to study expression differences among experimental groups of fruit flies, *Drosophila melanogaster*, subjected to various levels of inbreeding. They showed that inbreeding changed transcription levels for a number of genes. The genes that were differentially expressed were disproportionately involved in metabolism and stress responses, for example heat-shock proteins (*Hsp*), which are chaperones involved in folding and unfolding of intracellular proteins and macromolecules. Such genes are also upregulated during physiological stress and ageing. The results of this experiment suggest that inbreeding acts like an environmental stress factor.

As exemplified by this *Drosophila* study, microarray studies may be useful in studies relevant to conservation biology. However, microarrays are only available for a few organisms and although they may be constructed for any species it is not realistic that they will be developed for any but a few specialized studies (e.g. Vera *et al.* 2008). Microarrays developed for a model species may be tried in related threatened species (e.g. Abzhanov *et al.* 2006).

7.4.4 **Phylogenetics**

Genomic resources combined with phylogenetics provide insight in studies of phylogeographic patterns and intraspecific phylogenies. Such studies address whether recognized subspecies have been separated and if so for how long. A phylogenetic approach also provides better estimates of past effective population sizes (Edwards *et al.* 2007). These are important parameters for understanding local adaptation. How long and how much separation is needed for divergence among populations? These issues are of vital importance for inferring evolutionarily significant units and therefore defining management units. Multilocus genealogical approaches are still uncommon in phylogeography and historical demography, which have been dominated by microsatellite markers and chloroplast and mitochondrial DNA.

Theoretical studies of the coalescent process show that gene trees are not the same as species trees (Nichols 2001) and to estimate a species phylogeny the information from many genes are needed. For example, a research protocol was outlined by Liu and Pearl (2007) and Edwards *et al.* (2007) for estimating species trees as distinct from gene trees. In the so-called BEST approach, a Bayesian method for estimating species trees (Liu and Pearl 2007), vectors of gene trees are first estimated using a tree generated from a constrained set of preliminary species trees. These posterior distributions of gene tree vectors will provide the raw data for (maximum likelihood and) Bayesian estimation of phylogeny, population divergence times, and ancestral population sizes by exploring species tree space and maximizing the likelihood of gene tree vectors using the coalescent model of Rannala and Yang (2003). Ancestral population sizes can also be estimated with a method that uses gene tree–species tree conflicts in multilocus data sets (Nei 1987). Even incompletely resolved nuclear gene trees, when summed over multiple loci, can provide a strong signal for inference of demographic history (Jennings and Edwards 2005, Edwards *et al.* 2007).

With a subspecies tree and divergence time estimates it is possible to study how genes subjected to natural selection behave as compared with neutral gene and species trees. In the case of mammalian *Mhc* genes it has been observed that species share alleles that date back to before the split of the lineages (Edwards *et al.* 1995). The common explanation is that heterozygous genotypes are more fit, and hence extinction of alleles due to genetic drift is reduced drastically, resulting in long persistence times of alleles across speciation events. Long persistence times are also predicted by frequency-dependent models and other variants of balancing selection (Vekemans and Slatkin 1994). By contrast, some studies have found more divergence in *Mhc* alleles between populations than for neutral genes, suggesting strong divergent selection, perhaps as responses to different habitats or parasite faunas, between species (Miller *et al.* 1997, Ekblom *et al.* 2007, Saether *et al.* 2007).

7.5 Genomic studies of non-model species

Despite the difficulties in applying molecular genetics to ecologically well-known non-model species, a few studies have made exceptional advances in this field. In the following I review a few of these studies.

Bonin and coworkers (2006) used AFLPs to screen genetic variation along an altitude gradient of populations of the common frog, *Rana temporaria*, in France. Among a large set of markers they identified four that were more differentiated (higher F_{ST} values) than would be expected by random genetic drift (Fig. 7.3). They then subdivided their data into a neutral data set containing all the loci which behaved according to neutral expectations and a an altitude data set containing the outlier loci. When calculating phylogenetic relationships with the neutral data set, populations that were geographically close clustered together, as expected. However, with the altitude data, populations clustered according to altitude and not geographic distance. This strongly suggests that the genomic regions containing the AFLP sites in the altitude data have been subjected to similar forms of selection and may contain functional loci that have responded to selection for life at high altitudes. The genetic architecture of the common frog is, however, not established and no candidate functional loci have been suggested so far. Similar studies also using AFLPs have found outlier loci affected by selection on microgeographic differences among intertidal snails, *Littorina saxatilis* (Wilding *et al.* 2001, Butlin 2008).

In a series of papers, the genetic architecture behind morphological differences among benthic and marine forms of three-spined sticklebacks, *Gasterosteus aculeatus*, have been revealed (Peichel *et al.* 2004, Albert and Schluter 2004, Colosimo *et al.* 2004, Shapiro *et al.* 2004). Using a range of quantitative and molecular genetic techniques and experimental crosses, a complete linkage map has been published and the number of linkage groups equals the number of chromosomes in the species. With this map the researchers have shown that two important morphological differences among benthic and marine sticklebacks, armour plates and gill raker number, map to independent chromosomal regions and thus are controlled by different sets of genes. Furthermore, three aspects of skeletal armour are controlled by a only few chromosomal regions suggesting that only a few major genes play an important role in shaping these differences (Fig. 7.4).

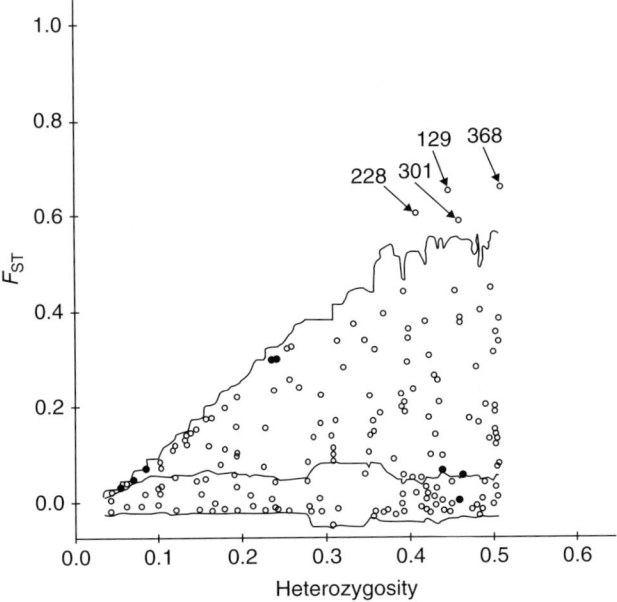

Figure 7.3 Plot of F_{ST} values against heterozygosity estimates comparing a high- and a low-altitude population. Each dot indicates an AFLP marker. The lower, intermediate, and higher lines represent the 5, 50, and 95% confidence intervals, respectively. Outlier loci are pointed out by arrows and referred to by numbers (from Bonin et al. 2006, reprinted with permission from the publisher)

Similarly, shape differences are best described by a geometric model of adaptation which states that a few major QTLs with a major effect are involved together with many genes with small pleiotropic effects. The major gene suggested to be involved and which maps to one of the markers with a major effect is the *EDA* gene. In humans 'the *EDA* gene provides instructions for making a protein called ectodysplasin A. This protein is part of a signalling pathway that plays an important role in development before birth. Specifically, it is critical for interactions between the ectoderm and the mesoderm embryonic cell layers. In the early embryo, these cell layers form the basis for many of the body's organs and tissues' (NIH Genetics Home reference; http://ghr.nlm.nih.gov/).

Another major difference between marine and benthic sticklebacks is the pelvic reduction that occurs in freshwater benthic forms. Previous studies had suggested that pelvic structures protect sticklebacks against gape-limited, soft-mouthed predators by presenting a lacerating defensive structure, increasing the effective diameter of the fish and resisting compressive forces during predator manipulation and chewing. However, several freshwater stickleback populations

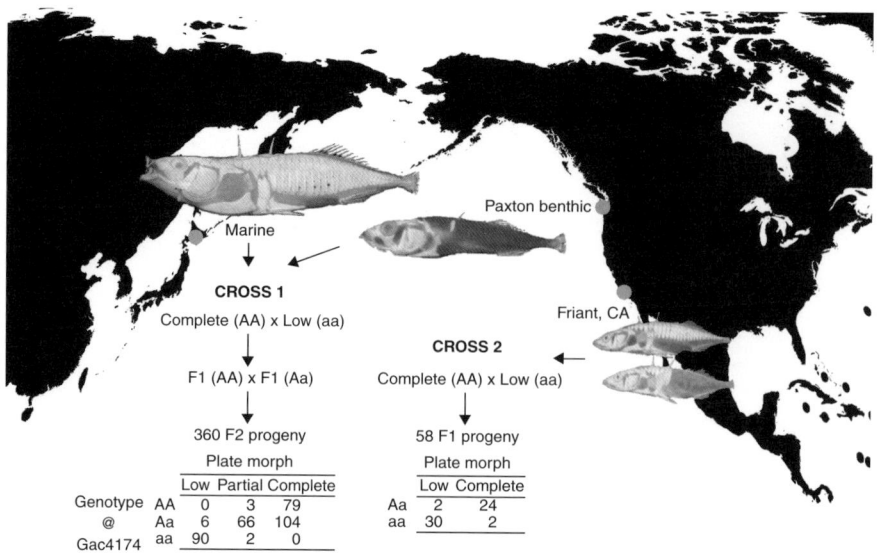

Figure 7.4 Mapping the genetic basis of lateral plate reduction in populations of three-spined sticklebacks. Dots show the geographic origins of the populations studied. AA, Aa, and aa refer to genotypes at Gac4174 (a microsatellite marker) near the major plate locus (Colosimo et al. 2004, reprinted with permission from the publisher).

have evolved complete or partial loss of the pelvic skeleton, perhaps in response to local absence of predatory fish (Shapiro et al. 2004). It was shown that the pelvic reduction was controlled by one major and four minor QTLs. The major gene involved was *Pitx1*, a gene expressed specifically during hindlimb development in mice, and which is required for normal hindlimb development in traditional vertebrate model systems. However, the sticklebacks did not show changes in Pitx1 protein sequence. Instead, pelvic-reduced sticklebacks showed site-specific regulatory changes in Pitx1 expression, with reduced or absent expression in pelvic and caudal fin precursors. It was thus suggested that regulatory mutations in major developmental control genes may provide a mechanism for generating rapid skeletal changes in natural populations, while preserving the essential roles of Pitx1 in other processes.

The adaptive radiation of the Darwin's finches is one of the textbook examples of how underlying differences in ecological conditions may cause divergent selection on beak morphologies and thus drive the speciation process in a group of birds of common origin (Lack 1947, Grant 1986). The genetic architecture of beak morphology have been studied in domestic chicken (Wu et al. 2004) and Abzhanov and coworkers (2004) studied one candidate gene, bone morphogenetic protein 4 (*Bmp4*), previously identified in chicken as a major gene involved in beak morphogenesis, in a set of species of Darwin's finches. By

Genomic studies of non-model species 137

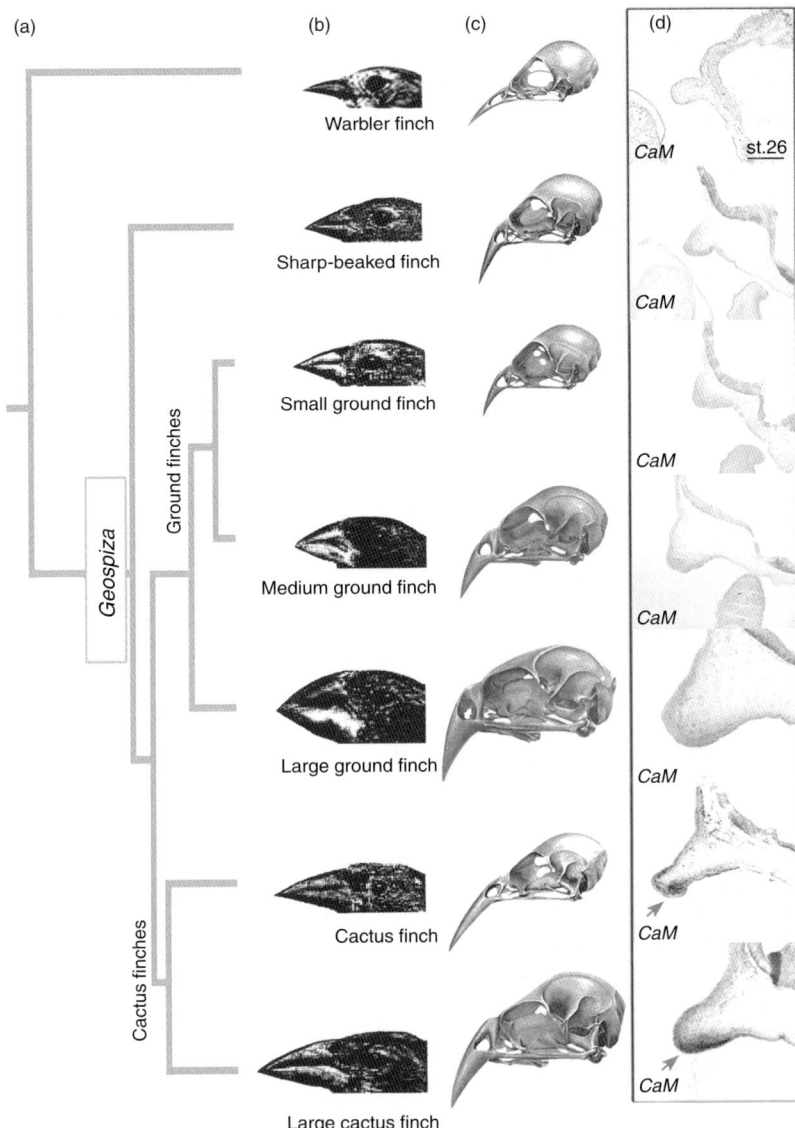

Figure 7.5 Phylogeny of *Geospiza* finches from the Galapagos islands with the different beak morphologies superimposed. The gene *CaM* is differentially expressed in the distal–ventral domain in the mesenchyme of the large-beaked species (from Abzhanov *et al.* 2006, reprinted with permission from the publisher).

performing comparative analysis of expression patterns in the different species, they identified variation in the level and timing of Bmp4 expression that correlated with variation in beak morphology. During development, Bmp4 is strongly expressed in a broad distal–dorsal domain in the mesenchyme of the upper beak prominence in species with large and broad beaks. The authors speculate that differences in the *cis*-regulatory elements of Bmp4 may underlie the distinct expression patterns found. This study is thus a classic example of the candidate gene approach (Box 7.1) in which a gene identified in a model (in this case the chicken) is shown also to be involved in the development of similar morphologies in non-model species.

In another study Abzhanov and coworkers used another approach to study the genetic background of differences in beak morphology in the same group of birds. By using a chicken microarray they were able to show that another protein, calmodulin (CaM), is differently expressed in Darwin's finches and that expression levels correlate with beak morphology (Fig. 7.5, Abzhanov *et al.* 2006). Calmodulin is a protein that binds and activates certain enzymes that trigger a signal which eventually turns specific genes on or off. Again this result suggests that regulatory genes and gene products play a major role in the evolution of divergent morphologies.

7.6 Conclusions

This chapter has reviewed the advances in genomics and their applications to conservation genetics. Tying back to Chapter 4 and the discussion on invasive species, Lee (2002) stated: 'the utility of genomic approaches for determining invasion mechanisms [are elucidated], through analysis of gene expression, gene interactions, and genomic rearrangements that are associated with invasion events.' She emphasized the utility of exploring genomic characteristics of invasive species, such as genes, gene complexes, and epistatic interactions, that promote invasive behaviour and concluded that such information could yield insights into the relationship between genetic architecture and rate of evolution, and evolutionary and ecological factors which confer invasion success. There is thus great hope in the new technologies opened up by the advances in genomics. Conservation and evolutionary biologists will now doubt lag behind the specialists working on model organisms, but the new techniques already have and will continue to have a major impact on studies in both conservation and evolution. Nevertheless, it is only by also studying adaptations in the field that we can gain a deeper understanding of how life forms have evolved and how we should best preserve biodiversity for future generations.

8 An evolutionary conservation biology

Humans have had and continue to have devastating effects on global biodiversity. It has been estimated that in just the last 400 years 127 named bird species have died out, all of them most probably due to human action (Newton 2003). These extinctions are not only actions of the modern industrialized culture, because as many as a thousand endemic island bird species may have disappeared following early human colonization in the pre-historic period (Milberg and Tyrberg 1993). Many extant bird species are at present critically endangered by human action. It is predicted that deforestation will cause just under 100 endemic bird species to become extinct on the islands of the Philippines and Indonesia alone in the near future (Brooks et al. 1996). Of existing tropical forests, 16 million ha are lost annually (Achard et al. 2002). It is forecasted that one in eight bird species may become extinct over the next 100 years worldwide (Sodhi et al. 2004). Nearly all (99%) of the threats are due to human activities such as deforestation and hunting (Butchart et al. 2004).

These numbers and bleak predictions are not just confined to birds but apply to all organisms. The rate of loss of populations and habitat for animal and plants is estimated to be about 1% per year (Balmford et al. 2003). On a regional scale in Britain, extinctions of butterflies, birds, and vascular plants were found to be correlated (Thomas et al. 2004). Thus the losses of birds mentioned in the previous paragraph are without doubt accompanied by losses in other taxa. There is also direct evidence that humans are impacting plants, too, as in Britain there is a relationship between the loss of scarce plants and human population density (Thompson and Jones 1999).

A recent summit on 'Evolutionary Change in Human-altered Environments' was hosted by the Institute of the Environment at the University of California in February 2007. In the report it is stated: 'As a consequence of [human-induced] impacts, we are witnessing a global, but unplanned, evolutionary experiment with the biotic diversity of the planet. Growing empirical evidence indicates that human-induced evolutionary changes impact every corner of the globe. Such changes are occurring rapidly, even at the level of a human lifespan, bear huge economical costs and pose serious threats to both humans and the biodiversity of the planet' (Smith and Bernatchez 2008). Humans have not only destroyed habitats and extinguished species, they have also changed species by domestication, moved them around the world, and released alien species into the wild. All these statements could also have been written in the present tense with an additional note to say that species are now not only transformed by traditional breeding but also by transgenic techniques. One of the major threats to human welfare is the spreading of resistance to antibiotics and pesticides among pests and disease organisms, a clear example of contemporary Darwinian evolution (Palumbi 2001). It is clear that humans have affected and are affecting the evolutionary process.

8.1 Human impact on evolutionary processes

As mentioned in Chapter 6, evolution was previously regarded as a slow process. However, evolutionary changes can occur within short periods of time (e.g. Reznick 1997, Hendry 2000, 2006, Bradshaw and Holzapfel 2001, Quinn 2001). Moreover, it has been recognized that what is primarily driving contemporary evolution are the same factors that are behind the present biodiversity crisis and the on-going extinction. The factors that have been identified as drivers of contemporary evolution are: habitat loss and fragmentation, overharvesting, and introduction of alien and invasive species (Stockwell *et al.* 2003). Rates of evolutionary change are measured in haldanes (in honour of one of the founders of modern evolutionary biology, J.B.S. Haldane), which is defined as standard deviations of phenotypic change per generation. Obviously the scale of change alters with time: the more generations that pass, the greater the change. Therefore, to judge whether a change has been faster or slower than expected, the residuals from this relationship need to be calculated (Fig. 8.1). It is obvious that evolutionary rates occur on timescales that can be observed and quantified but also that the range of evolutionary rates for different taxa over the same number of generations varies considerably.

Not all changes in morphology, behaviour, and life history observed in response to environmental perturbance (human-induced or not) are due to genetic microevolutionary change. For example, many saltwater species also occur in the brackish waters of the Baltic Sea and are thus able to reproduce and live at considerably lower salinities compared to that in which they are usually found (Johannesson and André 2006). Whether local Baltic populations have truly adapted to the lower salinity in the Baltic or if they represent species that are phenotypically plastic remains unclear.

Both microevolutionary change and phenotypic plasticity seem to matter for explaining occurrences of saltwater species in the Baltic. In flounder *Platichtys* possible for future change. It has long been recognized that practical conservation decisions should be based on evolutionary considerations. Erwin (1991) argued that, in a phylogeny of evolutionary lineages, the part in which species radiated the most should be given priority in protection. The part of the phylogeny in which rare endemics are found would be doomed to extinction in any case as such life forms obviously are not evolving. The radiating part of the phylogeny, on the other hand, represents evolutionary potential. If followed, this strategy would be different from many practical policies in which rare endemic species are considered conservation priorities. Arguments to preserve processes rather than patterns have also been made by several others (e.g. Smith *et al.* 1993, Thompson 1996, Stockwell *et al.* 2003)

A similar debate about conservation priorities prevails when it comes to prioritising among populations within species. Should conservation efforts and resources be put into large and thriving populations that have good future prospects or should they be put into small and peripheral populations? Lesica and Allendorf (1995) argued that the conservation value of peripheral populations depends upon their genetic divergence from other conspecific populations (Fig. 8.2). If peripheral populations are genetically and morphologically divergent from central populations such populations contribute to the overall genetic

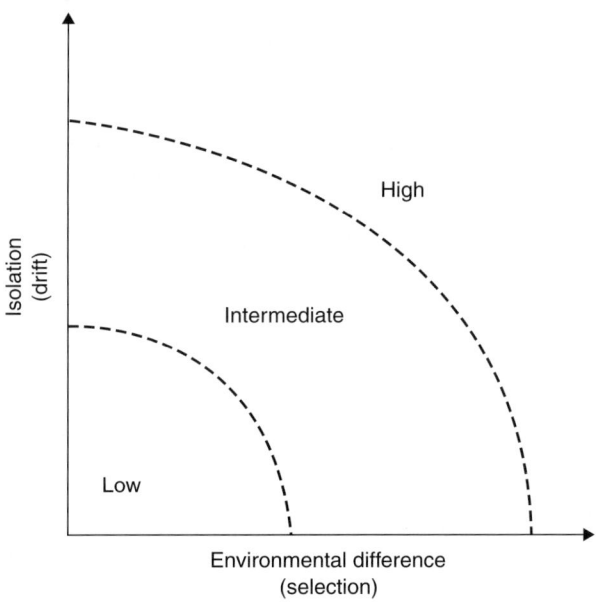

Figure 8.2 Relative conservation value of peripheral populations from an evolutionary perspective (from Lesica and Allendorf 1995, reprinted with permission from the publisher).

Table 8.1 Harvest-induced evolutionary changes in fish. For some stocks (n) the magnitude of change was quantified (from Jørgensen *et al.* 2007).

	No. of species	No. of studies	Change, % (n)
Maturation at lower age	6	10	23–24 (1)
Maturation at smaller size	7	13	20–33 (3)
Lower PMRN midpoint	5	10	3–49 (13)
Reduced annual growth	6	6	15–33 (3)
Increased fecundity	3	4	5–100 (3)
Loss of genetic diversity	3	3	21–22 (2)

PMRN, probabilistic maturation reaction norm.

large fish will be removed from the population while smaller and more difficult-to-catch fish are left behind (Lande *et al.* 1997).

A similar effect has been noted in hunted game populations. Trophy hunting selects against spectacular traits like large horns and antlers meaning that the hunted populations consist of fewer and fewer trophy animals (Coltman *et al.* 2003).

These evolutionary effects should come as no surprise. Ever since the dawn of evolutionary biology, predators have been believed to be important for shaping adaptations in prey (Reznick and Travis 1996, Swaddle and Lockwood 1998, Reznick *et al.* 2001). Humans, just like any top predator, impose selective pressures that have changed and will continue to change the properties of their prey populations.

Latta (2008) pointed out a corollary with human-imposed selective pressures. On the one hand humans select for undesirable features in some organisms, for example by inducing smaller and more uncatchable fish and game, resistant pests and disease organisms, and resilient weeds. On the other hand, human action has made it hard for us to change populations that we do want to alter. As discussed in Chapter 1, selection is a less potent evolutionary force in small and endangered populations and thus there are limits to the adaptive potential in such small populations, which cannot evolve despite our conservation efforts (Willi *et al.* 2006). It has therefore been proposed that a major focus of conservation biology should be to preserve the evolutionary potential of natural populations.

8.3 Conserving evolutionary potential

Conservation and evolution are to some extent a contradiction in terms. Evolution implies change and conservation implies no change. However, in this context conservation should be understood in terms of providing conditions making it

Due to the erection of hydroelectric power dams, the spawning waters of salmonid fishes have been affected. Several measures have been implemented to counteract the damage to fish stocks induced by these changes. They include the introduction of fish ladders so that migrating fish can pass the dams and several variants of supportive breeding using hatchery-reared fish. A growing body of evidence has shown that these altered selection regimes can result in genetic changes (Fleming *et al.* 2000, Hutchings and Fraser 2008, Waples *et al.* 2008).

In the River Dalälven in Sweden, large numbers of hatchery-produced trout have been raised and released for decades to compensate for the loss of natural reproduction caused by several hydroelectric power plants. This captive stock was founded using wild fish caught in the river. When comparing wild and hatchery-produced fish, phenotypic differences with a presumed genetic basis were observed (Petersson and Järvi 1993, 1995, Petersson *et al.* 1996). However, careful genetic studies using microsatellites and allozymes have shown that there was no genetic differentiation among the stocks (Palm *et al.* 2003a). Unfortunately, the experiments testing for phenotypic differences were not designed in a way that allowed for discrimination between phenotypic plasticity and genetic effects. However, an explanation for the differences among the stocks in morphology and behaviour may be that the observed phenotypic differences represent the actions of non-genetic maternal effects. Such may be mediated, for example, by egg-size differences among wild and hatchery-reared females (Jonsson *et al.* 1996).

Thus not all phenotypic changes observed in relation to human-induced changes are of genetic origin and instead represent plastic phenotypic responses. However, phenotypic plasticity also has a genetic component and may thus respond to selection (Via and Lande 1985, Stearns and Koella 1986, Scheiner 1993). It may be that human-induced changes select for species that have evolved the ability of phenotypic plasticity: species that are pre-adapted to live in stressed and unstable environments. On the other hand, it has been suggested that phenotypic plasticity is one of the traits that may be lost when fish become domesticated in hatcheries (Hutchings and Fraser 2008).

8.2 Evolutionary responses of harvesting

In harvested or managed populations evolutionary change induced by selective harvesting can be rapid and has been documented in a number of cases. It is predicted from life-history theory that increased mortality favours evolution towards earlier sexual maturation at smaller size. Commercial fishing that is selective with respect to size, maturity status, behaviour, or morphology has been shown to cause such shifts (Jørgensen *et al.* 2007; Table 8.1). Moreover, from the point of view of the harvester these changes cause undesired changes: easily caught

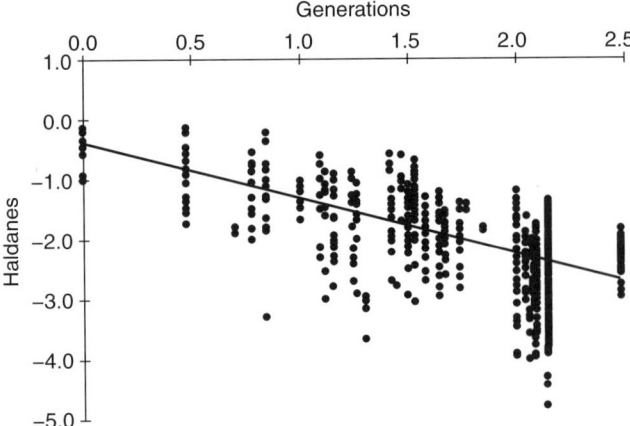

Figure 8.1 Evolutionary rates in haldanes relative to time interval, for a survey of 2104 rates. The trend line shows the mean predicted value over a given time interval (from Kinnison and Hendry 2001 and Stockwell *et al.* 2003, reprinted with permission from the publisher).

flesus there are two forms that differ in egg characteristics (Florin and Höglund 2008). One is a pelagic spawning form that lays small and buoyant eggs and the other shows demersal spawning with larger, more robust eggs that sink to the bottom. In the Baltic, salinities fall from the Sound to the east and northwards. The demersal spawning form is more common in the more brackish waters of the northern Baltic. The buoyant eggs of pelagic spawners cannot float in the lower salinities in the north Baltic and suffer great mortality. On the other hand, the sinking eggs of demersal spawners are more robust and can survive the mechanic forces on the bottom of the spawning banks. It is likely that this difference in spawning behaviour and egg characteristics represents a microevolutionary response to salinity.

The turbot, *Psetta maxima*, is another flatfish that occurs in marine waters and in the Baltic. Their ability to survive and reproduce at low salinity is more likely to be explained by phenotypic plasticity as we could find no population structure among fish caught on the saline west coast of Sweden and fish caught in the Baltic (Florin and Höglund 2007).

In a review of existing studies it was found that observed phenotypic changes were greater when the environmental change was anthropogenic than natural (Hendry *et al.* 2008). This difference may be explained by phenotypic plasticity rather than genetic change. In quantitative genetic studies that were designed to minimize the effects of phenotypic plasticity there was no difference among studies in which the change was anthropogenic or natural. However, the effect was evident for studies of wild-caught individuals in which both genetic and plastic responses may be present.

diversity within a species and would add to their long-term conservation. They argued further that peripheral populations are potentially important sites of future speciation events.

In California valley oak, *Quercus lobata*, some populations were more threatened than others. When such populations consisted of individuals with distinctive histories and genetic composition, they should be given priority in reserve network design or else valuable evolutionary information would be lost for this species (Grivet *et al.* 2008). In Eurasian populations of the nominate subspecies of the black-tailed godwit *Limosa limosa limosa*, unique and rare mitochondrial haplotypes were found in peripheral populations on the Baltic islands Öland and Gotland (Höglund *et al.* 2008). These islands harbour small fringe populations and yet mitochondrial diversity was much higher in these populations than in the much larger population breeding in the Netherlands. Clearly unique mitochondrial haplotypes would be lost if the Baltic populations should perish. Unfortunately, population-size trajectories have been negative in the past years and at the time of writing the Baltic populations are on the very brink of extinction.

What would be the optimal way to incorporate knowledge of evolutionary processes and the distribution of genetic diversity into conservation planning? Moritz (2002) argued for separation of genetic diversity into two dimensions, one concerned with adaptive variation and the other with neutral divergence caused by isolation. Conservation of species and specific areas should emphasize protection of historically isolated lineages or so-called evolutionarily significant units (ESUs) because these cannot be recovered. By contrast, adaptive features may best be protected by maintaining the context for selection, heterogeneous landscapes, and viable populations, rather than protecting specific phenotypes. Moritz proposed to (1) identify areas that are important for representing species and vicariant genetic diversity (by vicariant he meant genetic diversity specific to a particular area) and (2) within these areas maximize the protection of contiguous environmental gradients across which selection and migration can interact to maintain population viability and (adaptive) genetic diversity.

8.4 Conservation units

The need to identify conservation priorities has lead to the establishment of the concepts of ESUs (Ryder 1986) and management units (MUs; Moritz 1994), topics that have been touched upon previously in this book. How to best define such units is unclear (Crandall *et al.* 2000) and a heated debate on this topic used to prevail in the literature (see Fraser and Bernatchez 2001 for a review).

Many definitions of an ESU have been provided, each stressing different factors as important (Table 8.2). The ESU concept was first proposed to deal with the problems and vagueness of using subspecies definitions as a guide in conservation work. The original definition stressed that an ESU should be defined as a group of organisms that has been isolated from other conspecific groups for a sufficient period of time to have undergone meaningful genetic divergence from those other groups (Ryder 1986). In reality ESUs have been delimited by identifying groups of reciprocally monophyletic mitochondrial DNA lineages. Thus to qualify as an ESU, all lineages within a group must share a more recent common ancestor than any other lineage from another group (Moritz 1994).

Many conservation projects have also collected allele frequency data from allozymes and microsatellites. Such data were not easily applicable to the ESU concept and it was suggested that MUs could be used as a subcategory to ESUs. To qualify as MUs, populations should show significant differences in allele

Table 8.2 Evolutionarily significant unit (ESU) criteria (after Dyland and Bernatchez 2001).

Study	Criteria
Ryder 1986	Subsets of the more inclusive entity species, which possess genetic attributes significant for the present and future generations of the species in question
Waples 1991	A population or group of populations that:
	(i) is substantially reproductively isolated from other conspecific population units; and
	(ii) represents an important component of the evolutionary legacy of the species
Dizon et al. 1992	Populations or groups of populations demonstrating significant divergence in allele frequencies
Avise 1994	Sets of populations derived from consistently congruent gene phylogenies
Moritz 1994	Populations that:
	(i) are reciprocal monophyletic for mtDNA alleles; and
	(ii) demonstrate significant divergence of allele frequencies at nuclear loci
Vogler and DeSalle 1994	Groups that are diagnosed by characters which cluster individuals or populations to the exclusion of other such clusters
Crandall et al. 2000	Abandon the term ESU for a more holistic concept of species, consisting of populations with varying levels of gene flow evolving through drift and selection
Fraser and Bernatchez 2001	A lineage demonstrating highly restricted gene flow from other such lineages within the higher organizational level (lineage) of the species

distributions (Moritz 1994). In an ecological context an MU was defined as a group in which local population dynamics are determined primarily by birth and death rather than immigration and emigration (Moritz 1995).

The applicability of ESUs in situations where populations are continuously distributed has been questioned (Paetkau 1999). However, the main critique of the use of ESU is that when applied, the decision to call a species or population an ESU has most often been based on neutral characters. More genetic markers and the inclusion of non-neutral markers have been called for (Pertoldi *et al.* 2007).

This argument may be illustrated by a study of populations of the endangered North American Karner blue butterfly, *Lycaeides melissa samuelis* (Gompert *et al.* 2006). This subspecies is morphologically distinct from the nominate subspecies the Melissa blue butterfly, *Lycaeides melissa melissa*. It was shown that the presence of Melissa blue mitochondrial haplotypes in western Karner blue populations were the result of mitochondrial introgression. Thus western Karner blues were indistinct from Melissa blue butterflies on the basis of mtDNA whereas eastern populations were distinct. The subspecies were clearly separated in nuclear DNA which illustrates the risks of using data from a single locus for diagnosing ESUs.

The concept ESU has been important in practical management and legislation. In a review Fallon (2007) found that a taxonomic unit was much more likely to be included under the US Endangered Species Act if it had been assigned ESU status based on genetic data. Moreover, the type and amount of genetic data used was correlated with whether or not genetic distinction was discovered (the more and the better the data the more distinctions). The author called for guidelines for the evaluation of genetic information to list or delist organisms under the Endangered Species Act and advocated the use of multiple genetic markers.

In an attempt to reconcile the many views on ESU, Fraser and Bernatchez (2001) advocated what they called 'adaptive evolutionary conservation'. In this approach many differing criteria could be used alone or in combination depending on the situation to determine the conservation status of species and other taxonomic units. They argued that a rigid, universal definition of an ESU across all species may not be possible. Instead they concluded that the main conservation goal should be to preserve both evolutionary processes and the ecological viability of populations. This would be accomplished by maintaining as many populations within the species as possible so that the process of evolution will not be constrained. To my knowledge this approach has not been applied and ESUs are still in use although more markers, and also non-neutral ones, are used.

The debate over ESU may partly reflect which markers have been in fashion. In the early 1990s mtDNA and phylogenetic reconstructions dominated the scene. With the advent of microsatellites in the mid-1990s there was a call for using allele frequency differences and hence the MU was introduced. Now when selected markers have become more common there is a call to also include non-neutral information. Ironically, the debate over non-neutral versus neutral variation was a major impetus for the initiation of the current debate. The debate over ESUs also parallels the endless discussion of species concepts (Fraser and Bernatchez 2001). Advocates of the phylogenetic and related species concepts tend to favour ESU criteria based on historical and phylogenetic foundations while advocates of biological and similar species concepts have advocated the use of frequency differences and adaptive markers.

Now, with the advent of comparative data on whole genomes it has become clear that genomic variation is quite complex. Parts of the genome may be extremely conserved (e.g. coding genes) whereas other regions are more liable to change. Phylogenetic reconstruction of the evolutionary relationships between species works because lineage sorting by genetic drift makes species monophyletic over time. However, as has become evident in the debate over ESUs, incomplete lineage sorting has the consequence that closely related species may share gene sequences. As an example, in North American prairie grouse, the three species of the genus *Tympanuchus* all share mitochondrial haplotypes (Lucchini *et al.* 2001). Also, even in cases when conservationists are dealing with good taxonomic species, gene pools are not closed. Horizontal gene transfer occurs via viruses and other vectors. Gene trees are not the same as species trees (Pamilo and Nei 1988, Nichols 2001). To reconstruct the phylogeny and hence guide conservation decisions the information from many genes need to be considered.

8.5 Concluding remarks

How should a science of evolutionary conservation biology be framed? Pertoldi, Biljsma, and Loeschcke (2007) listed five problems affecting conservation genetics that should be addressed by future studies. I agree on four of these, which are listed here.

1   The lack of sufficient integration of the sub-disciplines of conservation genetics. Being a multidisciplinary and applied subject, conservation genetics is borrowing theory, techniques, and analytical tools from related subjects. Although much progress has been made with publication of books and journals devoted to the field, researchers may still have their background and some of their other research in nearby fields. Evolution is a unifying theory of biology and it should be apparent that evolutionary studies and evolutionary thinking has much to offer in conservation research and practice. To resolve this worry, there should be less focus on new techniques and markers and more focus on asking and resolving relevant questions.

2   Inferring selection by means of neutral markers. As has been argued in this book and elsewhere (Hedrick 2001, Gilligan *et al.* 2005), the correlation between molecular diversity (e.g. heterozygosity) and quantitative genetic variation (e.g. heritability) is weak and becomes even weaker in expanding or declining populations. Much current research is focused on finding the molecular basis for quantitative variation and while there is much optimism that these issues may be resolved with new genomic techniques there are theoretical limits to what can be gained. Fisher's fundamental theorem of natural selection (see Chapter 2) tells us that the heritability of fitness-related traits is transitory and generally low. One needs to be very lucky to detect quantitative trait nucleotides for fitness-related traits in natural populations.

3   Inferring population dynamics by means of neutral markers. As was reviewed in Chapter 4, there is quest for inferring population processes from genetic data. However, many different demographic scenarios may produce similar genetic footprints. There is clearly a need for more integration among metapopulation ecological theory and population genetics. It is also the case that geneticists, systematists, and ecologists have slightly different views on what they mean by a population (Waples and Gaggiotti 2006). Geneticists tend to stress units that are in Hardy–Weinberg and linkage equilibrium whereas ecologists may define populations as entities in which there is density-dependent mortality and reproduction.

4   Genetic consequences of increased environmental variability (the answer to which is actually integrated with the fifth problem listed by Pertoldi *et al.* (2007); that is, lack of ecological relevance). As has been discussed in Chapter 4, climate change is one of the major challenges of today. Climates in the future are not only going to be warmer but more fluctuations are also predicted. We do not yet have good knowledge of these effects on genetic variability and how populations are going to respond evolutionarily to more stochastic environments. The conservation genetic paradigm is that the more variation there is, the better. This can be illustrated using a metaphor: the more tools in the tool box, the more problems can be solved. However, is there an upper limit to how large the tool box should be? In other words, should conservation genetic projects ultimately always be aimed at preserving and restoring as much variation as possible? Theoretical considerations suggest that genetic variation actually lowers fitness under selection (Lande and Shannon 1996).

Ecology and genetics, good old-fashioned ecological genetics, will continue to cross foster each other's disciplines and both subjects are integral parts of the study of evolution. One area where ecology, genetics, and evolution come together is in understanding disease dynamics. A more complex understanding of immunogenetics—linking studies of disease, genetic variation, and demographic—declines is high on the list of future research priorities. It has become clear that pathogens are emerging and re-emerging as significant threats to wildlife and human health at an increasing rate (Acevedo-Whitehouse and Cunningham 2006). Infectious disease may well be the final causative agent which makes small and endangered populations go extinct. Infectious disease has had large effects on feral populations when a disease to which there is no resistance has been introduced, not least in our own species. Thus a fuller and deeper understanding of the ecology and evolution of disease and disease resistance is not only of academic interest but also of importance in practical conservation.

Conservation genetics can be seen as the effort to influence the evolutionary process in ways that enhance the persistence of population (Latta 2008). To do so we obviously first need to know about genetic variation in threatened species and much of the research throughout the history of the discipline has been aimed at studying and describing this variation. In Sweden, the Natural Environmental Protection Agency recently financed and published a survey of all genetic studies on wild plants and animals in the country (Andersson *et al.* 2007). Accompanying the report was an explicit proposal to the government on how to collect, store, and use such data in practical conservation in the future. International guidelines on how to set up and run such services have also been proposed (Schwartz *et al.* 2006). The time has come to implement these suggestions and to ask relevant questions about the evolutionary fate of endangered populations around the globe.

References

Abdelkrim, J., Pascal, M. *et al.* (2005) Island colonization and founder effects: the invasion of the Guadeloupe islands by ship rats (*Rattus rattus*). *Molecular Ecology* **14**: 2923–2931.

Abzhanov, A., Protas, M. *et al.* (2004) Bmp4 and morphological variation of beaks in Darwin's finches. *Science* **305**: 1462–1465.

Abzhanov, A., Kuo, W. *et al.* (2006) The calmodulin pathway and evolution of elongated beak morphology in Darwin's finches. *Nature* **442**: 563–567.

Acevedo-Whitehouse, K. and Cunningham, A.A. (2006) Is MHC enough for understanding wildlife immunogenetics? *Trends in Ecology and Evolution* **21**: 433–438.

Achard, F., Eva, H. *et al.* (2002) Determination of deforestation rates of the world's humid tropical forests. *Science* **297**: 999–1002.

Adams, M.D., Kelley, J.M. *et al.* (1991) Complementary DNA sequencing: expressed sequence tags and human genome project. *Science* **252**: 1651–1656.

Aguilar, A., Roemer, G. *et al.* (2004) High MHC diversity maintained by balancing selection in an otherwise genetically monomorphic mammal. *Proceedings of the National Academy of Sciences USA* **101**: 3490–3494.

Albert, A.Y.K. and Schluter, D. (2004) Reproductive character displacement of male stickleback mate preference: reinforcement or direct selection? *Evolution* **58**: 1099–1107.

Allendorf, F.W. and Ryman, N. (2002) The role of genetics in population viability analysis. In *Population Viability Analysis*, S.R. Beissinger and D.R. McCullough (eds), pp. 50–85. Chicago, IL: University of Chicago Press.

Al-Rabab'ah, M. and Williams, C. (2004) An ancient bottleneck in the Lost Pines of central Texas. *Molecular Ecology* **13**: 1075–1084.

Andersson, A., Andersson, S. *et al.* (2007) *Genetisk variation hos vilda växter och djur i Sverige*. Bromma: Naturvårdsverket.

Andersson, A.-C. (2004) *Postglacial Population History of the Common Shrew (Sorex araneus) in Fennoscandia*. Uppsala: Uppsala University.

Andersson, L. and Georges, M. (2004) Domestic animal genomics: deciphering the genetics of complex traits. *Nature Reviews Genetics* **5**: 202–212.

Apanius, V., Penn, D. *et al.* (1997) The nature of selection on the major histocompatibility complex. *Critical Reviews in Immunology* **17**: 179–224.

Arkush, K., Giese, A. *et al.* (2002) Resistance to three pathogens in the endangered winter-run chinook salomon (*Oncorhynchus tshawytscha*): effects of inbreeding and major histocompatiblity complex genotypes. *Canadian Journal of Fisheries and Aquatic Sciences* **59**: 966–975.

Armbruster, P. and Reed, D.H. (2005) Inbreeding depression in benign and stressful environments. *Heredity* **95**: 235–242.

Avise JC. 1994. *Molecular Markers, Natural History and Evolution.* Chapman & Hall, New York.

Babik, W., Durka, W. *et al.* (2005) Sequence diversity of the MHC DRB gene in the Eurasian beaver (*Castor fiber*). *Molecular Ecology* **14**: 4249–4257.

Babik, W., Pabijan, M. *et al.* (2008) Contrasting patterns of variation in MHC loci in the Alpine newt. *Molecular Ecology* **17**: 2339–2355.

Backström, N., Brandström, M. *et al.* (2006) Genetic mapping in a natural population of collared flycatchers (*Ficedula albicollis*): conserved synteny but gene order rearrangements on the avian Z chromosome. *Genetics* **174**: 377–386.

Bainbridge, M., Warren, R. *et al.* (2006) Analysis of the prostate cancer cell line LNCaP transcriptome using a sequencing-by-synthesis approach. *BMC Genomics* **7**: 246.

Balasubramanian, S., Sureshkumar, S. *et al.* (2006) The PHYTOCHROME C photoreceptor gene mediates natural variation in flowering and growth responses of *Arabidopsis thaliana*. *Nature Genetics* **38**: 711–715.

Balloux, F., Amos, W. *et al.* (2004) Does heterozygosity estimate inbreeding in real populations. *Molecular Ecology* **13**: 3021–3031.

Balmford, A., Green, R.E. *et al.* (2003) Measuring the changing state of nature. *Trends in Ecology and Evolution* **18**: 326–330.

Banta, J.A., Dole, J. *et al.* (2007) Evidence of local adaptation to coarse-grained environmental variation in *Arabidopsis thaliana*. *Evolution* **61**: 2419–2432.

Bateson, P. (1993) Optimal outbreeding. In *Mate Choice*, P. Bateson (ed.), pp. 257–277. Cambridge, Cambridge University Press.

Beacham, T., Candy, J. *et al.* (2001) Evaluation and application of microsatellite and major histocompatibility complex variation for stock identification of coho salmon in British Columbia. *Transactions of the American Fisheries Society* **130**: 1116–1155.

Beaumont, L. and Balding, D. (2004) Identifying adaptive genetic divergence among populations from genome scans. *Molecular Ecology* **13**: 969–980.

Beck, S. and Trowsdale, J. (1999) Sequence organisation of the class II region of the human MHC. *Immunological Reviews* **167**: 201–210.

Beebee, T. and Rowe, G. (2001) Application of genetic bottleneck testing to the investigation of amphibian declines: a case study with natterjack toads. *Conservation Biology* **15**: 266–270.

Belkhir, K., Borsa, P. *et al.* (1996– 2001) *GENETIX 4.02, Logiciel Sous Windows TM Pour la Génétique Des Populations.* Montpellier: Laboratoire Génome, Populations, Interactions, Université de Montpellier II.

Bellinger, M.R., Johnson, J.A. *et al.* (2003) Loss of genetic diversity in Wisconsin greater prairie chickens following a population bottleneck in Wisconsin, USA. *Conservation Biology* **17**: 717–724.

Bell-Pedersen, D., Cassone, V.M. et al. (2005) Circadian rhythms from multiple oscillators: lessons from diverse organisms. *Nature Review Genetics* **6**: 544–556.

Bensch, S., Åkesson, S. et al. (2002) The use of AFLP to find an informative SNP: genetic differences across a migratory divide in willow warblers. *Molecular Ecology* **11**: 2359–2366.

Berger, L., Speare, R. et al. (1998) Chytridiomycosis causes amphibian mortality associated with population declines in the rain forests of Australia and Central America. *Proceedings of the National Academy of Sciences USA* **95**: 9031–9036.

Bernatchez, L. and Landry, C. (2003) MHC studies in non-model vertebrates: what have we learned about natural selection in 15 years? *Journal of Evolutionary Biology* **16**: 363–377.

Bernier, G. and Périlleux, C. (2005) A physiological overview of the genetics of flowering time control. *Plant Biotechnology Journal* **3**: 3–16.

Bierne, N. and Eyre-Walker, A. (2003) The problem of counting sites in the estimation of the synonymous and nonsynonymous substitution rates: implications for the correlation between the synonymous substitution rate and codon usage bias. *Genetics* **165**: 1587–1597.

Bijlsma, R., Bundgaard, J. et al. (1999) Environmental dependence of inbreeding depression and purging in *Drosophila melanogaster*. *Journal of Evolutionary Biology* **12**: 1125–1137.

Bijlsma, R., Bundgaard, J. et al. (2000) Does inbreeding affect the extinction risk of small populations?: predictions from *Drosophila*. *Journal of Evolutionary Biology* **13**: 502–514.

BirdLife International (2004) Birds in Europe: population estimates, trends and conservation status. *BirdLife Conservation Series.* Barcelona: BirdLife International.

Bleeker, W. and Hurka, H. (2001) Introgressive hybridization in *Rorippa* (Brassicaceae): gene flow and its consequences in natural and anthropogenic habitats. *Molecular Ecology* **10**: 2013–2022.

Blondel, J., Dias, P.C. et al. (1993) Habitat heterogeneity and life-history variation of Mediterranean blue tits (*Parus caeruleus*). *The Auk* **110**: 511–520.

Boake, C.B. (2002) Sexual signaling and speciation, a microevolutionary perspective. *Genetica* **116**: 205–214.

Boerrigter, E.J.M. (1995) *On the Perspectives of Populations of the Rare Plant Species Phyteuma nigra*. PhD thesis, State University of Groningen, Groningen.

Bollmer, J., Vargas, F. et al. (2007) Low MHC variation in the endangered Galápagos penguin (*Spheniscus mendiculus*). *Immunogenetics* **59**: 593–602.

Bonin, A., Taberlet, P. et al. (2006) Explorative genome scan to detect candidate loci for adaptation along a gradient of altitude in the common frog (*Rana temporaria*). *Molecular Biology and Evolution* **23**: 773–783.

Bonneaud, C., Sorci, G. et al. (2004) Diversity of Mhc class I and IIB genes in house sparrows (*Passer domesticus*). *Immunogenetics* **55**: 855–865.

Bonneaud, C., Pérez-Tris, J. et al. (2006) Major histocompatibility alleles associated with local resistance to malaria in a passerine. *Evolution* **60**: 383–9.

Bonnell, M.L. and Selander, R.K. (1974) Elephant seals: genetic variation and near extinction. *Science* **184**: 908–909.

Bos, D. and DeWoody, J. (2005) Molecular characterization of major histocompatibility complex class II alleles in wild tiger salamanders (*Ambystoma tigrinum*). *Immunogenetics* **57**: 775–781.

Bouck, A. and Vision, T. (2007) The molecular ecologist's guide to expressed sequence tags. *Molecular Ecology* **16**: 907–924.

Bouzat, J.L., Cheng, H.H. et al. (1998a) Genetic evaluation of a demographic bottleneck in the greater prairie chicken. *Conservation Biology* **12**: 836–843.

Bouzat, J.L., Lewin, H.A. et al. (1998b) The ghost of genetic diversity past: historical DNA analysis of the greater prairie chicken. *American Naturalist* **152**: 1–6.

Bradshaw, W.W. and Holzapfel, C.M. (2001) Genetic shift in photoperiodic response correlated with global warming. *Proceedings of the National Academy of Sciences USA* **98**: 14509–14511.

Brooks, T., Pimm, S. et al. (1996) Deforestation predicts the number of threatened bird species in insular south east Asia. *Conservation Biology* **11**: 382–394.

Brown, J.H. and Kodrick-Brown, A. (1977) Turnover rates in insular biogeography: effect of immigration on extinction. *Ecology* **58**: 445–449.

Brown, J.L. and Eklund, A. (1994) Kin recognition and the major histocompatibility complex: an integrative review. *American Naturalist* **143**: 435–461.

Bryant, E.H., Meffert, L.M. et al. (1990) Fitness rebound in serially bottlenecked populations of the house fly. *American Naturalist* **114**: 1191–1211.

Bryant, E.H., Backus, V.L. et al. (1999) Experimental tests of captive breeding for endangered species. *Conservation Biology* **13**: 1487–1496.

Buckland, P. (2004) Allele-specific gene expression differences in humans. *Human Molecular Genetics* **13**: R255–R260.

Busch, J., Waser, P. et al. (2007) Recent demographic bottlenecks are not accompanied by a genetic signature in banner-tailed kangaroo rats (*Dipodomys spectabilis*). *Molecular Ecology* **16**: 2450–2462.

Butchart, S.H.M., Stattersfield, A.J. et al. (2004) Measuring global trends in the status of biodiversity: red list indices for birds. *PLoS Biology* **2**: e383.

Butlin, R.K. (2008) Population genomics and speciation. *Genetica* (in press).

Cain, A.J. and Sheppard, P.M. (1950) Selection in the polymorphic land snail *Cepaea nemoralis*. *Heredity* **4**: 275–294.

Cain, A.J. and Sheppard, P.M. (1954) The theory of adaptive polymorphism. *American Naturalist* **88**: 321–326.

Caizergues, A., Rätti, O. et al. (2003) Population genetic structure of male black grouse (*Tetrao tetrix*) in fragmented vs. continuous landscapes. *Molecular Ecology* **12**: 2297–2305.

Campos, J., Posada, D. et al. (2006) Genetic variation at MHC, mitochondrial and microsatellite loci in isolated populations of Brown trout (*Salmo trutta*). *Conservation Genetics* **7**: 515–530.

Carlsson, J. and Nilsson, J. (2001) Effects of geomorphological structures on genetic differentiation among brown trout populations in a northern boreal river drainage. *Transactions of the American Fisheries Society* **130**: 36–45.

Caro, T.M. and Laurenson, M.K. (1994) Ecological and genetic factors in conservation: a cautionary tale. *Science* **263**: 485–486.

Castric, V., Bernatchez, L. *et al.* (2001) Heterozygote deficiencies in small lacustrine populations of brook charr *Salvelinus fontinalis Mitchill* (Pisces, Salmonidae): a test of alternative hypotheses. *Heredity* **89**: 27–35.

Caughley, G. (1994) Directions in conservation biology. *Journal of Animal Ecology* **63**: 215–244.

Cavalli-Sforza, L.L., Moroni, A. *et al.* (2004) *Consanguinity, Inbreeding and Genetic Drift in Italy*. Princeton, NJ: Princeton University Press.

Cegelski, C., Waits, L. *et al.* (2003) Assessing population structure and gene flow in Montana wolverines (*Gulo gulo*) using assignment-based approaches. *Molecular Ecology* **12**: 2907–2918.

Charlesworth, B., Charlesworth, D. *et al.* (2003) The effects of genetic and geographic structure on neutral variation. *Annual Review of Ecology, Evolution, and Systematics* **34**: 99–125.

Charlesworth, D. and B. Charlesworth (1987) Inbreeding depression and its evolutionary consequences. *Annual Review of Ecology and Systematics* **18**: 237.

Charlesworth, D. and Charlesworth, B. (1999) The genetic basis of inbreeding depression. *Genetical Research* **74**: 329–340.

Cheung, F., Haas, B. *et al.* (2006) Sequencing *Medicago truncatula* expressed sequenced tags using 454 Life Sciences technology. *BMC Genomics* **7**: 272.

Cheverud, J. and Routman, E. (1996) Epistasis as a source of increased additive genetic variance of population bottleneck. *Evolution* **47**: 1042–1051.

Chistiakow, D.A., Hellemans, B. *et al.* (2006) Microsatellites and their genomic distribution, evolution, function and applications: a review with special reference to fish genetics. *Aquaculture* **255**: 1–29.

Cole, C.T. (2003) Genetic variation in rare and common plants. *Annual Review of Ecology Evolution and Systematics* **34**: 213–237.

Colosimo, P.F., Peichel, C.L. *et al.* (2004) The genetic architecture of parallel armor plate reduction in threespine sticklebacks. *PLoS Biology* **2**: e109.

Coltman, D., O'Donoghue, P. *et al.* (2003) Undesirable evolutionary consequences of trophy hunting. *Nature* **426**: 655–658.

Coltman, D.W. and Slate, J. (2003) Microsatellite measures of inbreeding: a meta analysis. *Evolution* **57**: 971–983.

Coltman, D.W., Bowen, W.D. *et al.* (1998) Birth weight and neonatal survival of harbour seal pups are positively correlated with genetic variation measured by microsatellites. *Proceedings of the Royal Society of London Series B Biological Sciences* **265**: 803–809.

Coltman, D.W., Pilkington, J.G. *et al.* (1999) Parasite-mediated selection against inbred Soay sheep in a free-living, island population. *Evolution* **53**: 1259–1267.

Conner, J.K. and Hartl, D.L. (2004) *A Primer of Ecological Genetics*. Sunderland, MA: Sinauer Associates.

Corander, J., Waldmann, P. *et al.* (2003) Bayesian analysis of genetic differentiation between populations. *Genetics* **163**: 367–374.

Corin, S.E., Abbott, K.L. *et al.* (2007) Large scale unicoloniality: the population and colony structure of the invasive Argentine ant (*Linepithema humile*) in New Zealand. *Insectes Sociaux* **54**: 275–282.

Cork, J.M. and Purugganan, M.D. (2005) High-diversity genes in the *Arabidopsis* genome. *Genetics* **170**: 1897–1911.

Cornuet, J. and Luikart, G. (1996) Description and power analysis of two tests for detecting recent population bottlenecks from allele frequency data. *Genetics* **144**: 2001–2014.

Cornuet, J., Piry, S. *et al.* (1999) New methods employing multilocus genotypes to select or exclude populations as origins of individuals. *Genetics* **153**: 1989–2000.

Costello, A.B., Down, T.E. *et al.* (2003) The influence of history and contemporary stream hydrology on the evolution of genetic diversity within species: an examination of microsatellite DNA variation in bull trout, *Salvelinus confluentus* (Pisces: Salmonidae). *Evolution* **57**: 328–344.

Coulon, A., Guillot, G. *et al.* (2006) Genetic structure is influenced by landscape features: empirical evidence from a roe deer population. *Molecular Ecology* **15**: 1669–1679.

Coulson, T.N., Pemberton, J.M. *et al.* (1998) Microsatellites reveal heterosis in red deer. *Proceedings of the Royal Society of London Series B Biological Sciences* **265**: 489–495.

Coulson, T.N., Albon, S.D. *et al.* (1999) Microsatellite loci reveal sex-dependent responses to inbreeding and outbreeding in red deer calves. *Evolution* **53**: 1951–1960.

Crandall, K.A., Bininda-Emonds, O.R.P. *et al.* (2000) Considering evolutionary processes in conservation biology. *Trends in Ecology and Evolution* **15**: 290–295.

Creel, S. (2006) Recovery of the Florida panther—genetic rescue, demographic rescue, or both? Response to Pimm *et al.* (2006). *Animal Conservation* **9**: 125–126.

Crnokrak, P. and Roff, D.A. (1999) Inbreeding depression in the wild. *Heredity* **83**: 260–270.

Culver, M., Hedrick, P.W. *et al.* (2008) Estimation of the bottleneck size in Florida panthers. *Animal Conservation* **11**: 104–110.

Dalén, L., Fuglei, E. *et al.* (2005) Population history and genetic structure of a circumpolar species: the arctic fox. *Biological Journal of the Linnean Society* **84**: 79–89.

Dannewitz, J., Maess, G. *et al.* (2005) Panmixia in the European eel: a matter of time. *Proceedings of the Royal Society of London Series B Biological Sciences* **272**: 1129–1137.

Daszak, P., Berger, L. *et al.* (1999) Emerging infectious diseases and amphibian population declines. *Emerging Infectious Diseases* **5**: 735–748.

Daszak, P., Cunningham, A.A. *et al.* (2003) Infectious disease and amphibian population declines. *Diversity and Distributions* **9**: 141–150.

Dawson, K. and Belkhir, K. (2001) A Bayesian approach to the identification of panmictic populations and the assignment of individuals. *Genetical Research* **78**: 59–77.

de Maynadier, P. and Hunter, M. (1999) Forest canopy closure and juvenile emigration by pond-breeding amphibians in Maine. *Journal of Wildlife Management* **63**: 441–450.

de Vienne, D., Bost, B. *et al.* (2001) Genetic variability of proteome expression and metabolic control. *Plant Physiology and Biochemistry* **39**: 271–283.

Dias, P. and Blondel, J. (1997) Local specialization and maladaptation in the Mediterranean blue tit (*Parus caeruleus*). *Oecologia* **107**: 79–86.

Dice, L.R. (1947) Effectiveness of selection by owls of deer-mice (*Peromyscus maniculatus*) which contrast in color with their background. *Contributions from the Laboratory of Vertebrate Zoology, University of Michigan* **34**: 1–20.

Dizon, A.E., Lockyer, C., Perrin, W.F. *et al.* (1992) Rethinking the stock concept: a phylogeographic approach. *Conservation Biology* **6**: 24–36.

Dobzhansky, T. (1948) Genetics of natural populations. Xviii. Experiments on chromosomes of *Drosophila pseudoobscura* from different geographic regions. *Genetics* **33**: 588–602.

Doerge, R. (2002) Mapping and analysis of quantitative trait loci in experimental populations. *Nature Reviews Genetics* **3**: 43–52.

Doherty, P.C. and Zinkernagel, R.M. (1975) A biological role for the major histocompatibility antigens. *Lancet* **1**: 1406–1409.

Dong, Q., Kroiss, L. *et al.* (2005) Comparative EST analysis in plant systems. *Methods in Enzymology: Producing the Biochemical Data* **395**: 400–418.

Drake, B.M., Goto, R.M. *et al.* (1999) Molecular and immunogenetic analysis of major histocompatibility haplotypes in northern bobwhite enable direct identification of corresponding haplotypes in an endangered subspecies, the masked bobwhite. *Zoo Biology* **18**: 279–294.

Easteal, S. (1981) The history of introductions of *Bufo marinus* (Amphibia: Anura): a natural experiment in evolution. *Biological Journal of the Linnean Society* **16**: 93–113.

Edwards, S., Wakeland, E. *et al.* (1995) Contrasting histories of avian and mammalian Mhc genes revealed by class II B sequences from songbirds. *Proceedings of the National Academy of Sciences, USA* **92**: 12200–12204.

Edwards, S., Gasper, J. *et al.* (1998) Genomics and polymorphism of Agph-DAB1, an Mhc class II B gene in red-winged blackbirds (*Agelaius phoeniceus*). *Molecular Biological Evolution* **15**: 236–250.

Edwards, S.V. and Hedrick, P.W. (1998) Evolution and ecology of MHC molecules: from genomics to sexual selection. *Trends in Ecology and Evolution* **13**: 305–311.

Edwards, S.V., Gasper, J. *et al.* (2000) A 39-kb sequence around a blackbird Mhc class II gene: ghost of selection past and songbird genome architecture. *Molecular Biological Evolution* **17**: 1384–1395.

Edwards, S.V., Liu, L. *et al.* (2007) High-resolution species trees without concatenation. *Proceedings of the National Academy of Sciences USA* **104**: 5936–5941.

Eisen, J. (1999) Mechanistic basis for microstallite instability. In *Microstallites, Evolution and Applications*, D. Goldstein and C. Schlötterer (eds), pp. 34–78. Oxford: Oxford University Press.

Ekblom, R. (2003) *Immunoecology of the Great Snipe (Gallinago media): Mate Choice, MHC Variation, and Humoral Immunocompetence in a Lekking Bird*. PhD thesis, Uppsala University, Uppsala.

Ekblom, R., Grahn, M. *et al.* (2003) Patterns of polymorphism in the MHC class II of a non-passerine bird, the great snipe (*Gallinago media*). *Immunogenetics* **54**: 734–41.

Ekblom, R., Saether, S.A. *et al.* (2007) Spatial pattern of MHC class II variation in the great snipe (*Gallinago media*). *Molecular Ecology* **16**: 1439–1451.

Ekblom, R., Saether, S. *et al.* (2008) Sexual selection, geographic structure and balancing selection in MHC genes of Great Snipe. *Genetica* in press.

Elgar, M.A. and Clode, D. (2001) Inbreeding and extinction in island populations: a cautionary note. *Conservation Biology* **15**: 284–286.

Ellegren, H. (1999) Inbreeding and relatedness in Scandinavian grey wolves *Canis lupus*. *Hereditas* **130**: 239–244.

Ellegren, H., Hartman, G. *et al.* (1993) Major histocompatibility complex monomorphism and low levels of DNA fingerprinting variability in a reintroduced and rapidly expanding population of beavers. *Proceedings of the National Academy of Sciences USA* **90**: 8150–8153.

Ellegren, H., Mikko, S. *et al.* (1996) Limited polymorphism at major histocompatibility complex (MHC) loci in the Swedish moose *A. alces*. *Molecular Ecology* **5**: 3–9.

El-Mousadik, A. and Petit, R.J. (1996) Chloroplast DNA phylogeography of the argan tree of Morocco. *Molecular Ecology* **5**: 547–555.

Epps, C., Palsbøll, P. *et al.* (2005) Highways block gene flow and cause a rapid decline in genetic diversity of desert bighorn sheep. *Ecology Letters* **8**: 1029–1038.

Erickson, D., Fenster, C. *et al.* (2004) Quantitative trait locus analyses and the study of evolutionary process. *Molecular Ecology* **13**: 2505–2522.

Erwin, D.H. (2006) *Extinction: How Life on Earth Nearly Ended 250 Million Years Ago*. Princeton, NJ: Princeton University Press.

Erwin, T. (1991) An evolutionary basis for conservation strategies. *Science* **253**: 750–753.

Estoup, A., Wilson, I.J. *et al.* (2001) Inferring population history from microsatellite and enzyme data in serially introduced cane toads, *Bufo marinus*. *Genetics* **159**: 1671–1687.

Estoup, A., Beaumont, M. *et al.* (2004) Genetic analysis of complex demographic scenarios: spatially expanding populations of the Cane toad, *Bufo marinus*. *Evolution* **58**: 2021–2036.

Evanno, G., Regnaut, S. *et al.* (2005) Detecting the number of clusters of individuals using the software STRUCTURE: a simulation study. *Molecular Ecology* **14**: 2611–2620.

Fagan, W.F. and Holmes, E.E. (2006) Quantifying the extinction vortex. *Ecology Letters* **9**: 51–60.

Falconer, D.S. and Mackay, T.F.C. (1996) *Introduction to Quantitative Genetics*. Harlow: Addison Wesley Longman.

Fallon, S.M. (2007) Genetic data and the listing of species under the U.S. Endangered Species Act. *Conservation Biology* **21**: 1186–1195.

Fay, J.C. and Wu, C.-I. (2000) Hitchhiking under positive Darwinian selection. *Genetics* **155**: 1405–1413.

Fenster, C.B. and Galloway, L.F. (2000) Inbreeding and outbreeding depression in natural populations of *Chamaecrista fasciculata* (Fabaceae). *Conservation Biology* **14**: 1406–1412.

Ferreira, A.G.A. and Amos, W. (2006) Inbreeding depression and multiple regions showing heterozygote advantage in *Drosophila melanogaster* exposed to stress. *Molecular Ecology* **15**: 3885–3893.

Ferrière, R., Dieckmann, U. *et al.* (eds) (2004) *Evolutionary Conservation Biology*. Cambridge: Cambridge University Press.

Fidler, A. and Gwinner, E. (2003) Comparative analysis of avian BMAL1 and CLOCK protein sequences: a search for features associated with owl nocturnal behaviour. *Comparative Biochemistry and Physiology Part B Biochemistry and Molecular Biology* **136**: 861–874.

Fisher, R.A. (1930) *The Genetical Theory of Natural Selection*. Oxford: Clarendon Press.

Flajnik, M., Ohta, Y. *et al.* (1999) Insight into the primordial MHC from studies in ectothermic vertebrates. *Immunological Reviews* **167**: 59–67.

Fleischmann, R.D., Adams, M.D. *et al.* (1995) Whole-genome random sequencing and assembly of *Haemophilus influenzae* Rd. *Science* **269**: 496–512.

Fleming, I., Hindar, K. *et al.* (2000) Lifetime success and interactions of farm salmon invading a native population. *Proceedings of the Royal Society of London Series B Biological Sciences* **267**: 1517–1523.

Flint, J. and Mott, R. (2001) Finding the molecular basis for quantitative traits: successes and pitfalls. *Nature Reviews Genetics* **2**: 437–445.

Florin, A.-B. and Höglund, J. (2007) Absence of population structure of turbot (*Psetta maxima*) in the Baltic Sea. *Molecular Ecology* **16**: 115–126.

Florin, A.-B. and Höglund, J. (2008) Population structure of flounder (*Platichthys flesus*) in the Baltic Sea: differences among demersal and pelagic spawners. *Heredity* **101**: 27–38.

Fondon, J. and Garner, H. (2004) Molecular origins of rapid and continuous morphological evolution. *Proceedings of the National Academy of Sciences USA* **101**: 18058–18063.

Ford, M. (2002) Applications of selective neutrality tests to molecular ecology. *Molecular Ecology* **11**: 1245–1262.

Frankel, O.H. and Soulé, M.E. (1981) *Conservation and Evolution*. New York: Cambridge University Press.

Frankham, R. (1995) Conservation genetics. *Annual Review of Genetics* **29**: 305–327.

Frankham, R. (1996) Relationship of genetic variation to populaton size in wildlife. *Conservation Biology* **10**: 1500–1508.

Frankham, R. (1999) Quantitative genetics in conservation biology. *Genetical Research* **74** 237–244.

Frankham, R., Gilligan, D.M. *et al.* (2001) Inbreeding and extinction: effects of purging. *Conservation Genetics* **2**: 279–284.

Frankham, R., Ballou, J.D. *et al.* (2002) *Introduction to Conservation Genetics*. Cambridge: Cambridge University Press.

Franklin, I.R. (1980) Evolutionary change in small populations. In *Conservation Biology*, M.E. Soulé and M. Wilcox (eds), pp. 137–149. Sunderland, MA: Sinauer Associates.

Fraser, D.J. and Bernatchez, L. (2001) Adaptive evolutionary conservation: towards a unified concept for defining conservation units. *Molecular Ecology* **10**: 2741–2752.

Freeman-Gallant, C., Johnson, E. *et al.* (2002) Variation at the major histocompatibility complex in savannah sparrows. *Molecular Ecology* **11**: 1125–1130.

Frentiu, F., Clegg, S. et al. (2008) Pedigree-free animal models: the relatedness matrix reloaded. *Proceedings of the Royal Society of London Series B Biological Sciences* **275**: 639–647.

Fu, Y. (1996) New statistical tests of neutrality for DNA samples from a population. *Genetics* **143**: 557–570.

Fu, Y.X. and Li, W.H. (1993) Statistical tests of neutrality of mutations. *Genetics* **133**: 693–709.

Funk, W., Blouin, M. et al. (2005) Population structure of Columbia spotted frogs (*Rana luteiventris*) is strongly affected by the landscape. *Molecular Ecology* **14**: 483–496.

Gaggiotti, O.E. (2003) Genetic threats to population persistence. *Annales Zoologici Fennici* **40**: 155–168.

Gamfeldt, L. and Källström, B. (2007) Increasing intraspecific diversity increases predictability in population survival in the face of perturbations. *Oikos* **116**: 700–705.

Garrigan, D. and Hedrick, P. (2003) Detecting adaptive molecular polymorphism: lessons from the mhc. *Evolution* **57**: 1707–1722.

Garza, J. and Williamson, E. (2001) Detection of reduction in population size using data from microsatellite loci. *Molecular Ecology* **10**: 305–318.

Gibbs, H.L. and Grant, P.R. (1989) Inbreeding in Darwin's medium ground finches (*Geospiza fortis*). *Evolution* **43**: 1273–1284.

Gilligan, D.M., Briscoe, D.A. et al. (2005) Comparative losses of quantitative and molecular genetic variation in finite populations of *Drosophila melanogaster*. *Genetics Research* **85**: 47–55.

Gilpin, M.E. and Soulé, M.E. (1986) Minimum viable populations: processes of species extinctions. In *Conservation Biology: the Science of Scarcity and Diversity*, M. Soulé (ed.), pp. 19–34. Sunderland, MA: Sinauer Associates.

Goldstein, D.B. and Schlötterer, C. (eds) (1999) *Microsatellites, Evolution and Applications*. Oxford: Oxford University Press.

Gompert, Z., Nice, C.C. et al. (2006) Identifying units for conservation using molecular systematics: the cautionary tale of the Karner blue butterfly. *Molecular Ecology* **15**: 1759–1768.

Goossens, B., Chikhi, L. et al. (2006) Genetic signature of anthropogenic population collapse in orang-utans. *PLoS Biology* **4**: e25.

Gorg, A., Weiss, W. et al. (2004) Current two-dimensional electrophoresis technology for proteomics. *Proteomics* **4**: 3665–3685.

Goto, R., Afanassieff, M. et al. (2002) Single-strand conformation polymorphism (SSCP) assays for major histocompatibility complex B genotyping in chickens. *Poultry Science* **81**: 1832–1841.

Goudet, J. (2002) *FSTAT version 2.9.3.2*. Lausanne: University of Lausanne.

Goudet, J. and Büchi, L. (2006) The effects of dominance, regular inbreeding and sampling design on QST, an estimator of population differentiation for quantitative traits. *Genetics* **172**: 1337–1347.

Grahn, M. and Forsberg, L. (2008) Homozygote advantage at the MHC: more evidence foroutbreeding depression. In *Genetic Aspects of Sexual Selection and Mate Choice in Salmonids*, L. Forsberg (ed.). Uppsala: Uppsala University.

Grant, B.R. and Grant, P.R. (1989) *Evolutionary Dynamics of a Natural Population: the Large Cactus Finch of the Galápagos*. Chicago, IL: University of Chicago Press.

Grant, P.R. (1986) *Ecology and Evolution of Darwin's Finches*. Princeton, NJ: Princeton University Press.

Grant, P.R. and Grant, B.R. (1995) The founding of a new population of Darwin's finches. *Evolution* **49**: 229–240.

Grant, P.R. and Grant, B.R. (2002) Unpredictable evolution in a 30-year study of Darwin's finches. *Science* **296**: 707–711.

Grant, P.R. and Grant, B.R. (2007) *How and Why Species Multiply: the Radiation of Darwin's Finches*. Princeton, NJ: Princeton University Press.

Grimholt, U., Getahun, A. *et al.* (2000) The major histocompatibility class II alpha chain in salmonid fishes. *Developmental & Comparative Immunology* **24**: 751–763.

Grivet, D., Sork, V.L. *et al.* (2008) Conserving the evolutionary potential of California valley oak (*Quercus lobata* Née): a multivariate genetic approach to conservation planning. *Molecular Ecology* **17**: 139–156.

Groombridge, J.J., Jones, C.G. *et al.* (2000) Conservation biology: 'ghost' alleles of the Mauritius kestrel. *Nature* **403**: 616.

Gross, A.O. (1928) The heath hen. *Memoirs of the Boston Society of Natural History* **6**: 489–588.

Gross, A.O. (1930) *Progress Report of the Wisconsin Prairie Chicken Investigation*. Madison, WI: Wisconsin Conservation Commission.

Guillot, G., Estoup, A. *et al.* (2005) A spatial statistical model for landscape genetics. *Genetics* **170**: 1261–1280.

Gustafsson, L. (1986) Lifetime reproductive success and heritability: empirical support for Fisher's fundamental theorem. *American Naturalist* **128**: 761–764.

Gustafsson, L. (1988) Interspecific competition lowers fitness in the collared flycatcher (*Ficedula albicollis*). *Ecology* **68**: 281–286.

Gutierrez-Espeleta, G.A., Hedrick, P.W. *et al.* (2001) Is the decline of desert bighorn sheep from infectious disease the result of low MHC variation? *Heredity* **86**: 439–450.

Gyllenstrand, N., Clapham, D. *et al.* (2007) A Norway spruce FLOWERING LOCUS T homolog is implicated in control of growth rhythm in conifers. *Plant Physiol.* **144**: 248–257.

Hajji, G., Zachos, F. *et al.* (2007) Conservation genetics of the imperilled Barbary red deer in Tunisia. *Animal Conservation* **10**: 229–235.

Hampe, A. and Petit, R. (2005) Conserving biodiversity under climate change: the rear edge matters. *Ecology Letters* **8**: 461–467.

Hansson, B. and Westerberg, L. (2002) On the correlation between heterozygosity and fitness in natural populations. *Molecular Ecology* **11**: 2467–2474.

Hansson, B., Åkesson, M. *et al.* (2005) Linkage mapping reveals sex-dimorphic map distances in a passerine bird. *Proceedings of the Royal Society of London Series B Biological Sciences* **272**: 2289–2298.

Hassel, K., Såstad, S.M. *et al.* (2005) Genetic variation and structure in the expanding moss *Pogonatum dentatum* (Polytrichaceae) in its area of origin and in a recently colonized area. *American Journal of Botany* **92**: 1684–1690.

Hauswaldt, J., Stuckas, H. *et al.* (2007) Molecular characterization of MHC class II in a nonmodel anuran species, the fire-bellied toad *Bombina bombina*. *Immunogenetics* **59**: 479–491.

Hedrick, P. (1999) Perspective: highly variable loci and their interpretation in evolution and conservation. *Evolution* **53**: 313–318.

Hedrick, P. (2005) A standardized genetic differentiation measure. *Evolution* **59**: 1633–1638.

Hedrick, P.W. (2006) Genetic polymorphism in heterogeneous environments: the age of genomics. *Annual Review of Ecology, Evolution and Systematics* **37**: 67–93.

Hedrick, P., Parker, K. *et al.* (2000) Major histocompatibility complex variation in the Arabian oryx. *Evolution* **54**: 2145–2151.

Hedrick, P.W. (2000) *Genetics of Populations*. Sudbury, MA: Jones and Bartlet.

Hedrick, P.W. (2001) Conservation genetics: where are we now? *Trends in Ecology and Evolution* **16**: 629–636.

Hedrick, P.W. (2002) Pathogen resistance and genetic variation at MHC loci. *Evolution* **56**: 1902–1908.

Hedrick, P.W. and Kalinowski, S.T. (2000) Inbreeding depression in conservation biology. *Annual review of Ecology and Systematics* **31**: 139–216.

Hendry, A. (2000) Rapid evolution of reproductive isolation in the wild: evidence from introduced salmon. *Science* **290**: 516–518.

Hendry, A.P., Grant, P.R. *et al.* (2006) Possible human impacts on adaptive radiation: beak size bimodality in Darwin's finches. *Proceedings of the Royal Society of London Series B Biological Sciences* **273**: 1887–1894.

Hendry, A.P., Farrugia, T.J. *et al.* (2008) Human influences on rates of phenotypic change in wild animal populations. *Molecular Ecology* **17**: 20–29.

Hewitt, G.M., Butlin, R.K. *et al.* (1987) Testicular dysfunction in hybrids between parapatric subspecies of the grasshopper *Chorthippus parallelus*. *Biological Journal of the Linnean Society* **31**: 25–34.

Hirschhorn, J. and Daly, M. (2005) Genome-wide association studies for common diseases and complex traits. *Nature Reviews Genetics* **6**: 95–108.

Hitchings, S.P. and Beebee, T.J.C. (1998) Loss of genetic diversity and fitness in common toad (*Bufo bufo*) populations isolated by inimical habitat. *Journal of Evolutionary Biology* **11**: 269–283.

Hoekstra, H. (2006) Genetics, development and evolution of adaptive pigmentation in vertebrates. *Heredity* **97**: 222–234.

Hoekstra, H. and Nachman, M. (2003) Different genes underlie adaptive melanism in different populations of rock pocket mice. *Molecular Ecology* **12**: 1185–1194.

Hoekstra, H., Krenz, J. et al. (2005) Local adaptation in the rock pocket mouse (*Chaeotdipus intermedius*): natural selection and phylogenetic history of populations. *Heredity* **94**: 217–228.

Hoelzel, A., Stephens, J. et al. (1999) Molecular genetic diversity and evolution at the MHC DQB locus in four species of pinnipeds. *Molecular Biology and Evolution* **16**: 611–618.

Hoelzel, A.R., Halley, J. et al. (1993) Elephant seal genetic variation and the use of simulation models to investigate historical population bottlenecks. *Journal of Heredity* **84**: 443–449.

Hofman, M. (2007) *The Effect of Genetic Diversity on Evolutionary Potential: Local Adaptation in the Natterjack Toad Bufo calamita in Sweden*. MSc thesis, Uppsala, Uppsala University.

Höglund, J. and Shorey, L. (2003) Local genetic structure in a white-bearded manakin population. *Molecular Ecology* **12**: 2457–2463.

Höglund, J. and Shorey, L. (2004) Genetic divergence in the superspecies Manacus. *Biological Journal of the Linnean Society* **81**: 439–447.

Höglund, J., Piertney, S.B. et al. (2002) Inbreeding and male fitness in a wild population. *Proceedings of the Royal Society of London Series B Biological Sciences* **269**: 711–715.

Höglund, J., Larsson, J.K. et al. (2007) Genetic variability in European black grouse (*Tetrao tetrix*). *Conservation Genetics* **8**: 239–243.

Höglund, J., Johansson, T. et al. (2008) Phylogeography of the black-tailed godwit *Limosa limosa*: substructuring revealed by mtDNA control region sequences. *Journal of Ornithology* (in press).

Höglund, L. (1961) The reactions of fish in concentration gradients. *Reprints from the Institute of Freshwater Research Drottningholm* **43**: 1–147.

Holderegger, R. and Wagner, H. (2006) A brief guide to landscape genetics. *Landscape Ecology* **21**: 793–796.

Holsinger, K.E., Mason-Gamer, R.J. et al. (1999) Genes, demes, and plant conservation. In *Genes, Species, and the Threat of Extinction: DNA and Genetics in the Conservation of Endangered Species*, L.F. Landweber and A.P. Dobson (eds), pp. 23–46. Princeton, NJ: Princeton University Press.

Horton, R., Wilming, L. et al. (2004) Gene map of the extended human MHC. *Nature Reviews Genetics* **5**: 889–899.

Houle, D. (1992) Comparing evolvability and variability of quantitative traits. *Genetics* **130**: 195–204.

Hudson, P. and Baines, D. (1993) Black grouse. In *The New Atlas of Breeding Birds in Britain and Ireland 1988–1991*, D. Gibbons et al. (eds), pp. 130–131. T. & A.D. Poyser, London.

Hudson, R., Kreitman, M. et al. (1987) A test of neutralmolecular evolution based on nucleotide data. *Genetics* **116**: 153–159.

Huennecke, L.F. (1991) Ecological implications of genetic variation in plant populations. In *Genetics and Conservation of Rare Plants*, D.A. Falk and K.E. Holsinger (eds), pp. 31–44. New York: Oxford University Press.

Hughes, A. and Nei, M. (1988) Pattern of nucleotide substitution at major histocompatibility complex class-I loci reveals overdominant selection. *Nature* **335**: 167–170.

Hunt, H., Goto, R. *et al.* (2006) At least one YMHCI molecule in the chicken is alloimmunogenic and dynamically expressed on spleen cells during development. *Immunogenetics* **58**: 297–307.

Hutchings, J.A. and Fraser, D.J. (2008) The nature of fisheries- and farming induced evolution. *Molecular Ecology* **17**: 294–313.

Ingvarsson, P.K. (2001) Restoration of genetic variation lost—the genetic rescue hypothesis. *Trends in Ecology and Evolution* **16**: 62–63.

Ingvarsson, P.K. (2002) Lone wolf to the rescue. *Nature* **420**: 472.

Ingvarsson, P.K. and Whitlock, M.C. (2000) Heterosis increases the effective migration rate. *Proceedings of the Royal Society of London Series B Biological Sciences* **267**: 1321–1326.

Ingvarsson, P.K., Garcia, M.V. *et al.* (2008) Nucleotide polymorphism and phenotypic associations within and around the phytochrome B2 locus in European aspen (*Populus tremula*, Salicaceae). *Genetics* **178**: 2217–2226.

IPCC (2007) *Climate Change- Synthesis Report*. Geneva: IPCC.

Jackson, I. (1994) Molecular and developmental genetics of mouse coat color. *Annual Review of Genetics* **28** 189–217.

Jansen, R. and Nap, J. (2001) Genetical genomics: the added value from segregation. *Trends in Genetics* **17**: 388–391.

Janssens, X., Fontaine, M. *et al.* (2008) Genetic pattern of the recent recovery of European otters in southern France. *Ecography* **31**: 176–186.

Jarvi, S.I., Tarr, C.L. *et al.* (2004) Natural selection of the major histocompatibility complex (MHC) in Hawaiian honeycreepers (Drepanidinae). *Molecular Ecology* **13**: 2157–68.

Jarvi, S.I., Tarr, C.L. *et al.* (2004) Natural selection of the major histocompatibility complex (Mhc) in Hawaiian honeycreepers (Drepanidinae). *Molecular Ecology* **13**: 2157–2168.

Jennings, W. andEdwards, S. (2005) Speciational history of Australian grass finches (Poephila) inferred from thirty gene trees. *Evolution* **59**: 2033–2047.

Jensen, L., Hansen, M. *et al.* (2008) Spatially and temporally fluctuating selection at non-MHC immune genes: evidence from TAP polymorphism in populations of brown trout (*Salmo trutta*, L.). *Heredity* **100**: 79–91.

Johannesson, K. and André, C. (2006) Life on the margin: genetic isolation and diversity loss in a peripheral marine ecosystem, the Baltic Sea. *Molecular Ecology* **15**: 2013–2029.

Johannesson, K., Johannesson, B. *et al.* (1995) Strong naturalselection causes microscale allozyme variation in a marine snail. *Proceedings of the National Academy of Sciences USA* **92**: 2602–2606.

Johansson, M., Primmer, C.R. *et al.* (2007) Does habitat fragmentation reduce fitness and adaptability? A case study of the common frog (*Rana temporaria*). *Molecular Ecology* **16**: 2693–2700.

Johnsen, A., Fidler, A.E. *et al.* (2007) Avian Clock gene polymorphism: evidence for a latitudinal cline in allele frequencies. *Molecular Ecology* **16**: 4867–4880.

Johnson, J.A. and Dunn, P.O. (2006) Low genetic variation in the heath hen prior to extinction and implications for the conservation of prairie-chicken populations. *Conservation Genetics* **7**: 37–48.

Jonsson, N., Jonsson, B. *et al.* (1996) Does early growth cause a phenotypically plastic response in egg production of Atlantic salmon? *Functional Ecology* **10**: 89–96.

Jordan, W.C. and Bruford, M.W. (1998) New perspectives on mate choice and the MHC. *Heredity* **81**: 127–133.

Jorde, P.E. and Ryman, N. (1995) Temporal allele frequency change and estimation of effective size in populations with overlapping generations. *Genetics* **139**: 1077–1090.

Jorde, P.E. and Ryman, N. (1996) Demographic genetics of brown trout (*Salmo trutta*) and estimation of effective population size from temporal change of allele frequencies. *Genetics* **143**: 1369–1381.

Jorde, P.E. and Ryman, N. (2007) Unbiased estimator for genetic drift and effective population size. *Genetics* **177**: 927–935.

Jørgensen, C., Enberg, K. *et al.* (2007) Managing evolving fish stocks. *Science* **318**: 1247–1248.

Kalinowski, S.T. and Hedrick, P.W. (1998) An improved method for estimating inbreeding depression in pedigrees. *Zoo Biology* **17**: 481–497.

Karp, A., Edwards, K.J. *et al.* (1997) Molecular technologies for biodiversity evaluation: opportunities and challenges. *Nature Biotechnology* **15**: 625–628.

Kauer, M., Dieringer, D. *et al.* (2003) Nonneutral admixture of immigrant genotypes in african *Drosophila melanogaster* populations from Zimbabwe. *Molecular Biology and Evolution* **20**: 1329–1337.

Kaufman, J. (2000) The simple chicken major histocompatibility complex: life and death in the face of pathogens and vaccines. *Philosophical Transactions of the Royal Society London Series B Biological Sciences* **355**: 1077–1084.

Kaufman, J., Milne, S. *et al.* (1999) The chicken B locus is a minimal essential major histocompatibility complex. *Nature* **401**: 923–925.

Kawamoto, Y., Tomari, K. *et al.* (2007) Genetics of the Shimokita macaque population suggest an ancient bottleneck. *Primates* **49**: 32–40.

Keller, L.F. (1998) Inbreeding and its fitness effects in an insular population of song sparrows (*Melospiza melodia*). *Evolution* **52**: 240–250.

Keller, L.F., Marr, A.B. *et al.* (2006) The genetic consequences of small population size:inbreeding and loss of genetic variation. In *Conservation and Biology of Small Populations*, J.N.M. Smith *et al.* (eds), pp. 113–137. New York: Oxford University Press.

Kempenaers, B. (2007) Mate choice and genetic quality: a review of the heterozygosity theory. *Advances in the Study of Behavior* **37**: 189–278.

Kerje, S., Lind, J. *et al.* (2003) Melanocortin 1-receptor (MC1R) mutations are associated with plumage colour in chicken. *Animal Genetics* **34**: 241–248.

Kettlewell, H.B.D. (1955) Recognition of appropriate backgrounds by the pale and black phases of Lepidoptera. *Nature* **175**: 943–944.

Kettlewell, H.B.D. (1956) A resume of investigations on the evolution of melanism in the Lepidoptera. *Proceedings of the Royal Society of London Series B Biological Sciences* **145**: 297–303.

Kettlewell, H.B.D. (1958) A survey of the frequencies of *Biston betularia* (L.) (Lep.) and its melanic forms in Great Britain. *Heredity* **12**: 51–72.

Kimura, M. (1968) Evolutionary rate at the molecular level. *Nature* **217**: 624–626.

Kimura, M. (1969) The number of heterozygous nucleotide sites maintained in a finite population due to steady flux of mutations. *Genetics* **61**: 893–903.

Kimura, M. (1983) *The Neutral Theory of Molecular Evolution*. Cambridge: Cambridge University Press.

Kimura, M. and Crow, J.F. (1963) The measurement of effective population number. *Evolution* **17**: 279–288.

Kinnison, M.T. and Hendry, A.P. (2001) The pace of modern life II: from rates of contemporary microevolution to pattern and process. *Genetica* **112–113**: 145–164.

Kirkpatrick, M. and Barton, N. (1997) Evolution of species's range. *American Naturalist* **150**: 1–23.

Kirkpatrick, M. and Jarne, P. (2000) The effects of a bottleneck on inbreeding depression and the genetic load. *American Naturalist* **155**: 154–167.

Klein, J., Sato, A. *et al.* (1998) Molecular trans-species polymorphism. *Annual Review of Ecology and Systematics* **29**: 1–21.

Knight, J. (2004) Allele-specific gene expression uncovered. *Trends in Genetics* **20**: 113–116.

Kolbe, J.J., Glor, R.E. *et al.* (2004) Genetic variation increases during biological invasion by a Cuban lizard. *Nature* **431**: 177–181.

Koskinen, M.T., Piironen, J. *et al.* (2001) Interpopulation genetic divergence in European grayling (*Thymallus thymallus*, Salmonidae) at a microgeographic scale: implications for conservation. *Conservation Genetics* **2**: 133–143.

Koskinen, M.T., Haugen, T.O. *et al.* (2002) Contemporary fisherian life-history evolution in small salmonid populations. *Nature* **419**: 826–830.

Kreitman, M. (2000) Methods to detect selection in populations with applications to the human. *Annual Review of Genomics & Human Genetics* **1**: 539–559.

Kreitman, M. and Akashi, H. (1995) Molecular evidence for natural selection. *Annual Review of Ecology and Systematics* **26**: 403–422.

Kristensen, T., Sørensen, P. *et al.* (2005) Genome-wide analysis on inbreeding effects on gene expression in *Drosophila melanogaster*. *Genetics* **171**: 157–167.

Krutovskii, K.V. and Neale, D.B. (2001) *Forest Genomics for Conserving Adaptive Genetic Diversity*. Forest Genetic Resources Working Papers. Forestry Department working paper FGR/3 (E). Forest Resources Development Service, Forest Resources Division. Rome: FAO.

Kruuk, L.E.B., Sheldon, B.C. *et al.* (2002) Severe inbreeding depression in collared flycatchers (*Ficedula albicollis*). *Proceedings of the Royal Society of London Series B Biological Sciences* **269**: 1581–1589.

Lack, D. (1947) *Darwin's Finches*. Cambridge: Cambridge University Press.
Lacy, R.C. and Ballou, J.D. (1998) Effectiveness of selection in reducing the genetic load in populations of *Peromyscus polionotus* during generations of inbreeding. *Evolution* **52**: 900–909.
Laikre, L. (1999) Conservation genetic management of brown trout (*Salmo trutta*) in Europe. *Report by the Concerted Action on Identification, Management and Exploitation of Genetic Resources in the Brown Trout ('TROUTCONCERT')*.
Laikre, L. and Ryman, N. (1991) Inbreeding depression in a captive wolf (*Canis lupus*) population. *Conservation Biology* **5**: 33–40.
Laikre, L. and Ryman, N. (1996) Effects on intraspecific biodiversity from harvesting and enhancing natural populations. *Ambio* **25**: 504–509.
Laikre, L., Ryman, N. *et al.* (1997) Estimated inbreeding in a small, wild muskox *Ovibus moschatus* population and its possible effects on population reproduction. *Biological Conservation* **19**: 197–204.
Lande, R. (1976) Natural selection and random genetic drift in phenotypic evolution. *Evolution* **30**: 314–334.
Lande, R. (1988) Genetics and demography in biological conservation. *Science* **241**: 1455–1460.
Lande, R. (1992) Neutral theory of quantitative genetic variance in an island model with local extinctions and recolonization. *Evolution* **46**: 381–389.
Lande, R. (1995) Mutation and conservation. *Conservation Biology* **9**: 782–791.
Lande, R. (1999) Extinction risks from anthropogenic, ecological and genetic factors. In *Genetics and the Extinction of Species*, L.F. Landweber and A.P. Dobson (eds), pp. 1–22. Princeton, NJ: Princeton University Press.
Lande, R. and Barrowclough, G.F. (1987) Effective populations size, genetic variation, and their use in population management. In *Viable Populations for Conservation*, M. Soulé (ed.), pp. 84–124. Cambridge: Cambridge University Press.
Lande, R. and Shannon, S. (1996) The role of genetic variation in adaptation and population persistence in a changing environment. *Evolution* **50**: 434–437.
Lande, R., Saether, B. *et al.* (1997) Threshold harvesting for sustainability of fluctuating resources. *Ecology* **78**: 1341–1350.
Landry, C. and Bernatchez, L. (2001) Comparative analysis of population structure across environments and geographical scales at major histocompatibility complex and microsatellite loci in Atlantic salmon (*Salmo salar*). *Molecular Ecology* **10**: 2525–2539.
Landweber, L. and Dobson, A. (eds) (1999) *Genetics and the Extinction of Species*. Princeton, NJ: Princeton University Press.
Langefors, A., Lohm, J. *et al.* (2001) Association between major histocombatibility complex class IIB alleles and resistance to *Aeromonas salmonicida* in Atlantic salmon. *Proceedings of the Royal Society of London Series B Biological Sciences* **268**: 479–485.
Larsson, J.K., Jansman, H.A.H. *et al.* (2008) Genetic impoverishment of the last black grouse (*Tetrao tetrix*) population in the Netherlands: detectable only with a reference from the past. *Molecular Ecology* **17**: 1897–1904.

Latta, R.G. (2008) Conservation genetics as applied evolution: from genetic pattern to evolutionary process. *Evolutionary Applications* **1**: 84–94.

Latter, B.D.H., Mulley, J.C. et al. (1995) Reduced genetic load revealed by slow inbreeding in *Drosophila melanogaster*. *Genetics* **139**: 287–297.

Laugen, A.T., Laurila, A. et al. (2003) Latitudinal countergradient variation in the common frog (*Rana temporaria*) development rates—evidence for local adaptation. *Journal of Evolutionary Biology* **16**: 996–1005.

Laurens, V., Chapusot, C. et al. (2001) Axolotl MHC class II beta chain: predominance of one allele and alternative splicing of the beta 1 domain. *European Journal of Immunology* **31**: 506–515.

Leder, E., Danzmann, R. et al. (2006) The candidate gene, Clock, localizes to a strong spawning time quantitative trait locus region in rainbow trout. *Journal of Heredity* **97**: 74–80.

Lee, C.E. (2002) Evolutionary genetics of invasive species. *Trends in Ecology and Evolution* **17**: 386–391.

Legendre, P. and Legendre, L. (1998) *Numerical Ecology*. Amsterdam: Elsevier.

Leinonen, T., O'Hara, R.B. et al. (2008) Comparative studies of quantitative trait and neutral marker divergence: a meta-analysis. *Journal of Evolutionary Biology* **21**: 1–17.

Lenormand, T. (2002) Gene flow and the limits to natural selection. *Trends in Ecology and Evolution* **17**: 183–189.

Lesica, P. and Allendorf, F. (1995) When are peripheral populations valuable for conservation? *Conservation Biology* **9**: 753–760.

Lever, C. (2001) *The Cane Toad. The History and Ecology of a Successful Colonist*. Otley, West Yorkshire: Westbury Academic and Scientific Publishing.

Lewontin, R.C. and Krakauer, J. (1973) Distribution of gene frequency as a test of the theory of selective neutrality of polymorphisms. *Genetics* **74**: 175–195.

Liberg, O.A., Andrén, H. et al. (2005) Severe inbreeding depression in a wild wolf (*Canis lupus*) population. *Biology Letters* **1**: 17–20.

Liu, L. and Pearl, D.K. (2007) Species trees from gene trees: reconstructing Bayesian posterior distributions of a species phylogeny using estimated gene tree distributions. *Systematic Biology* **56**: 504–514.

Loeschcke, V., Tomiuk, J. et al. (eds) (1994) *Conservation Genetics*. Basel: Birkhäuser Verlag.

Lönn, M., Prentice, H.C. et al. (1996) Genetic structure, allozyme-habitat associations and reproductive fitness in *Gypsophila fastigiata* (Caryophyllaceae). *Oecologia* **106**: 308–316.

Lönn, M. and Prentice, H.C. (2002) Gene diversity and demographic turnover in central and peripheral populations of the perennial herb *Gypsophila fastigiata*. *Oikos* **99**: 489–498.

Losos, J.B. (1990) Ecomorphology, performance capability, and scaling of West Indian *Anolis* lizards: an evolutionary analysis. *Ecological Monographs* **60**: 369–388.

Lucchini, V., Höglund, J. et al. (2001) Historical biogeography and a mitochondrial DNA phylogeny of grouse and ptarmigan. *Molecular Phylogenetics and Evolution* **20**: 149–162.

Lugon-Moulin, N. and Hausser, J. (2002) Phylogeographical structure, postglacial recolonization and barriers to gene flow in the distinctive Valais chromosome race of the common shrew (*Sorex araneus*). *Molecular Ecology* **11**: 785–794.

Luijten, S.H. (2001) *Reproduction and Genetics of Fragmented Plant Populations*. PhD thesis, University of Amsterdam, Amsterdam.

Luijten, S.H., Dierick, A. *et al*. (2000) Population size, genetic variation, and reproductive success in a rapidly declining, self-incompatible perennial (*Arnica montana*) in The Netherlands. *Conservation Biology* **14**: 1776–1787.

Luikart, G. and Cornuet, J. (1998) Empirical evaluation of a test for identifying recently bottlenecked populations from allele frequency data. *Conservation Biology* **12**: 228–237.

Luikart, G., England, P.R. *et al*. (2003) The power and promise of population genomics: from genotyping to genome typing. *Nature Review Genetics* **4**: 981–994.

Lynch, M., Burger, R. *et al*. (1993) The mutational meltdown in asexual populations. *Heredity* **47**: 1744–1757.

Lynch, M. and Milligan, B.G. (1994) Analysis of population genetic structure with RAPD markers. *Molecular Ecology* **3**: 91–99.

Lynch, M., Pfrender, M. *et al*. (1999) The quantitative and molecular genetic architecture of a subdivided species. *Evolution* **53**: 100–110.

Lyons, L., Laughlin, T. *et al*. (1997) Comparative anchor tagged sequences (CATS) for integrative mapping of mammalian genomes. *Nature Genetics* **15**: 47–56.

Madison, D. and Farrand, L. (1998) Habitat use during breeding and emigration in radio implanted tiger salamanders, *Ambystoma tigrinum*. *Copeia* **1998**: 402–410.

Madsen, T., Shrine, R. *et al*. (1999) Restoration of an inbred adder population. *Nature* **402**: 34–35.

Madsen, T., Olsson, M. *et al*. (2000) Population size and genetic diversity in sand lizards (*Lacerta agilis*) and adders (*Vipera berus*). *Biological Conservation* **94**: 257–262.

Madsen, T., Ujvari, B. *et al*. (2004) Novel genes continue to enhance population growth in adders (*Vipera berus*). *Biological Conservation* **120**: 145–147.

Maehr, D.S., Crowley, P. *et al*. (2006) Of cats and Haruspices: genetic intervention in the Florida panther. Response to Pimm *et al*. (2006). *Animal Conservation* **9**: 127–132.

Majerus, M.E.N. (1998) *Melanism: Evolution in Action*. Oxford: Oxford University Press.

Makova, K. and Norton, H. (2005) Worldwide polymorphism at the MC1R locus and normal pigmentation variation in humans. *Peptides* **26**: 1901–1908.

Malécot, G. (1948) *Mathématiques de l'hérédité*. Paris: Masson.

Mandak, B., Bimova, K. *et al*. (2005) Loss of genetic variation in geographically marginal populations of *Atriplex tatarica* (Chenopodiaceae). *Annals of Botany* **96**: 901–912.

Manel, S., Schwartz, M. *et al*. (2003) Landscape genetics: combining landscape ecology and population genetics. *Trends in Ecology and Evolution* **18**: 189–197.

Margulies, M., Egholm, M. *et al*. (2005) Genome sequencing in microfabricated high-density picolitre reactors. *Nature* **437**: 376–380.

Mariette, S., Le Corre, V. *et al*. (2002) Sampling within the genome for measuring within-population diversity: trade-offs between markers. *Molecular Ecology* **11**: 1145–1156.

Markert, J.A., Grant, P.R. et al. (2004) Neutral locus heterozygosity, inbreeding, and survival in Darwin's ground finches (*Geospiza fortis* and *G. scandens*). *Heredity* **92**: 306–315.

Maruyama, T. and Fuerst, P. (1984) Population bottlenecks and nonequilibrium models in population genetics. I. Allele numbers when populations evolve from zero variability. *Genetics* **108**: 745–763.

Maruyama, T. and Fuerst, P.A. (1985) Population bottlenecks and nonequilibrium models in population genetics. II. Number of alleles in a small population that was formed by a recent bottleneck. *Genetics* **111**: 675–689.

Massingham, T. and Goldman, N. (2005) Detecting amino acid sites under positive selection and purifying selection. *Genetics* **169**: 1753–1762.

Maynard Smith, J. (1998) *Evolutionary Genetics*. Oxford: Oxford University Press.

Mays, H.L. and Hill, G.E. (2004) Choosing mates: good genes versus genes that are a good fit. *Trends in Ecology and Evolution* **19**: 554–559.

McDonald, J. and Kreitman, M. (1991) Adaptive protein evolution at the Adh locus in Drosophila. *Nature* **351**: 652–654.

McKay, J.K. and Latta, R.G. (2002) Adaptive population divergence: markers, QTL and traits. *Trends in Ecology and Evolution* **17**: 285–291.

McKeigue, P. (2005) Prospects for admixture mapping of complex traits. *American Journal of Human Genetics* **76**: 1–7.

McNeilly, T. (1968) Evolution in closely adjacent plant populations. III. *Agrostis tenuis* on a small copper mine. *Heredity* **23**: 99–108.

McNielly, T. and Bradshaw, A.D. (1968) Evolutionary processes in populations of copper tolerant *Agrostis tenuis*. *Evolution* **22**: 108–18.

McRae, B., Beier, P. et al. (2005) Habitat barriers limit gene flow and illuminate historical events in a wide-ranging carnivore, the American puma. *Molecular Ecology* **14**: 1965–1977.

Meagher, S., Penn, D.J. et al. (2000) Male-male competition magnifies inbreeding depression in wild house mice. *Proceedings of the National Academy of Sciences USA* **97**: 3324–3329.

Merilä, J. and Sheldon, B.C. (2000) Lifetime reproductive success and heritability in nature. *American Naturalist* **155**: 301.

Merilä, J. and Crnokrak, P. (2001) Comparison of genetic differentiation at marker loci and quantitative traits. *Journal of Evolutionary Biology* **14**: 892–903.

Meyers, L.A. and Bull, J.J. (2002) Fighting change with change: adaptive variation in an uncertain world. *Trends in Ecology and Evolution* **17**: 551–557.

Mikko, S. and Andersson, L. (1995) Low major histocompatiblity complex class II diversity in European and North American moose. *Proceedings of the National Academy of Sciences USA* **92**: 4259–4263.

Milberg, P. and Tyrberg, T. (1993) Naive birds and noble savages - a review of man-caused prehistoric extinctions of island birds. *Ecography* **16**: 229–250.

Milinski, M. (2006) The major histocompatibility complex, sexual selection, and mate choice. *Annual Review of Ecology, Evolution, and Systematics* **37**: 159–186.

Miller, H.C. and Lambert, D.M. (2004) Gene duplication and gene conversion in class II MHC genes of New Zealand robins (Petroicidae). *Immunogenetics* **56**: 178–191.

Miller, K., Withler, R. *et al.* (1997) Molecular evolution at Mhc genes in two populations of chinook salmon *Onchorynchus tshawytscha*. *Molecular Ecology* **6**: 937–954.

Miller, M., Goto, R. *et al.* (1996) Assignment of Rfp-Y to the chicken major histocompatibility complex NOR microchromosome and evidence for high-frequency recombination associated with the nucleolar organizer region. *Proceedings of the National Academy of Sciences USA* **93**: 3958–3962.

Miller, M., Bacon, L. *et al.* (2004) Nomenclature for the chicken major histocompatibility (B and Y) complex. *Immunogenetics* **56**: 261–279.

Miller, M.M., Wang, C. *et al.* (2005) Characterization of two avian MHC-like genes reveals an ancient origin of the CD1 family. *Proceedings of the National Academy of Sciences USA* **102**: 8674–8679.

Mills, M.G.L. (2006) Response to article: 'The genetic rescue of the Florida panther' by Pimm *et al.* (2006). *Animal Conservation* **9**: 123–124.

Mishima, K., Tozawa, T. *et al.* (2005) The 3111T/C polymorphism of hClock is associated with evening preference and delayed sleep timing in a Japanese population sample. *American Journal of Medical Genetics Part B Neuropsychiatric Genetics* **133B**: 101–104.

Montalvo, A.M. and Ellstrand, N.C. (2000) Transplantation of the subshrub *Lotus scoparius*: testing the home-site advantage hypothesis. *Conservation Biology* **14**: 1034–1045.

Mooney, H. and Hobbs, R. (eds) (2000) *Invasive Species in a Changing World*. Washington: Island Press.

Mooney, H.A. and Cleland, E.E. (2001) The evolutionary impact of invasive species. *Proceedings of the National Academy of Sciences USA* **98**: 5446–5451.

Morin, P.A., Luikart, G. *et al.* (2004) SNPs in ecology, evolution, and conservation. *Trends in Ecology and Evolution* **19**: 208–216.

Moritz, C. (1994) Defining 'evolutionary significant units' for conservation. *Trends in Ecology and Evolution* **9**: 373–375.

Moritz, C. (1995) Uses of molecular phylogenies for conservation. *Philosophical Transactions of the Royal Society London Series B Biological Sciences* **349**: 113–118.

Moritz, C. (2002) Strategies to protect biological diversity and the evolutionary processes that sustain it. *Systematic Biology* **51**: 238–254.

Morton, N.E., Crow, J.F. *et al.* (1956) An estimate of the mutational damage in man from data on consanguineous marriages. *Proceedings of the National Academy of Sciences USA* **42**: 855–863.

Mundy, N. (2006) Genetic basis of color variation in wild birds. In *Bird Coloration*, G. Hill and K. McGraw (eds), pp. 469–505. Cambridge, MA: Harvard University Press.

Nadeau, N.J., Burke, T. *et al.* (2007a) Evolution of an avian pigmentation gene correlates with a measure of sexual selection. *Proceedings of the Royal Society of London Series B Biological Sciences* **274**.

Nadeau, N.J., Mundy, N.I. *et al.* (2007b) Association of a single-nucleotide substitution in TYRP1 with roux in Japanese quail (*Coturnix japonica*). *Animal Genetics* **38**: 609–613.

Nakadate, M., Shikano, T. *et al.* (2003) Inbreeding depression and heterosis in various quantitative traits of the guppy *Poecilia reticulata*. *Aquaculture* **220**: 219–226.

Nei, M. (1975) *Molecular Population Genetics and Evolution*. Amsterdam: North Holland.

Nei, M. (1987) *Molecular Evolutionary Genetics*. New York: Columbia University Press.

Nei, M. and Li, W.H. (1979) Mathematical model for studying genetic variation in terms of restriction endonucleases. *Proceedings of the National Academy of Sciences USA* **76**: 5269–5273.

Nei, M. and Tajima, F. (1981) DNA polymorphism detectable by restriction endonucleases. *Genetics* **97**: 146–163.

Nei, M. and Kumar, S. (2000) *Molecular Evolution and Phylogenetics*. New York: Oxford University Press.

Newman, D. and Pilson, D. (1997) Increased probability of extinction due to decreased genetic effective population size: experimental populations of *Clarkia pulchella*. *Evolution* **51**: 354–362.

Newton, I. (2003) *The Speciation and Biogeography of Birds*. New York: Academic Press.

Nichols, R. (2001) Gene trees and species trees are not the same. *Trends in Ecology and Evolution* **16**: 358–364.

Nielsen, R. (1997) A likelihood approach to populations samples of microsatellite alleles. *Genetics* **146**: 711–716.

Nielsen, R. and Yang, Z. (1998) Likelihood models for detecting positively selected amino acid sites and applications to the HIV-1 envelope gene. *Genetics* **148**: 929–936.

Nowak, M., Tarczy-Hornoch, K. *et al.* (1992) The optimal number of major histocompatibility complex molecules in an individual. *Proceedings of the National Academy of Sciences USA* **89**: 10896–10899.

Ober, C. (1999) Studies of HLA, fertility and mate choice in a human isolate. *Human Reproduction Update* **5**: 103–107.

O'Brien, S., Wildt, D. *et al.* (1985) The cheetah is depauperate in genetic variation. *Science* **221**: 459–462.

O'Hara, R.B. and Merilä, J. (2005) Bias and precision in Q_{ST} estimates: problems and some solutions. *Genetics* **171**: 1331–1339.

Ohta, Y., Goetz, W. *et al.* (2006) Ancestral organization of the MHC revealed in the amphibian *Xenopus*. *Journal of Immunology* **176**: 3674–3685.

Oostermeijer, J.G.B., van Eijck, M.W. *et al.* (1994) Offspring fitness in relation to population size and genetic variation in the rare perennial plant species *Gentiana pneumonanthe* (Gentianaceae). *Oecologia* **97**: 289–296.

Oostermeijer, J.G.B., van Eijck, M.W. *et al.* (1995) Analysis of the relationship between allozyme heterozygosity and fitness in the rare *Gentiana pneumonanthe* L. *Journal of Evolutionary Biology* **8**: 739–759.

Oostermeijer, J.G.B., Luijten, S.H. *et al.* (2003) Integrating demographic and genetic approaches in plant conservation. *Biological Conservation* **113**: 389–398.

Orr, H. and Coyne, J. (1992) The genetics of adaptation: a reassessment. *American Naturalist* **140**: 725–742.

Österberg, M.K., Shavorskaya, O. *et al.* (2002) Naturally occurring indel variation in the *Brassica nigra* COL1 gene is associated with variation in flowering time. *Genetics* **161**: 299–306.

Otto, S. (2000) Detecting the form of selection from DNA sequence data. *Trends in Genetics* **16**: 526–529.

Ovaskainen, O., Cano, J.M. *et al.* (2008) A Bayesian framework for comparative quantitative genetics. *Proceedings of the Royal Society of London Series B Biological Sciences* **275**: 669–678.

Paetkau, D. (1999) Using genetics to identify intraspecific conservation units: a critique of current methods. *Conservation Biology* **13**: 1507–1509.

Paetkau, D., Calvert, W. *et al.* (1995) Microsatellite analysis of population structure in Canadian polar bears. *Molecular Ecology* **4**: 347–354.

Paetkau, D., Shields, G. *et al.* (1998) Gene flow between insular, coastal and interior populations of brown bears in Alaska. *Molecular Ecology* **7**: 1283–1292.

Pagel, M. and Pomiankowski, A. (2007) The organismal prospect. In *Evolutionary Genomics and Proteomics*, M. Pagel and A. Pomiankowski (eds). Sunderland, MA, Sinauer Associates.

Palm, S., Dannewitz, J. *et al.* (2003a) Lack of molecular genetic divergence between sea-ranched and wild sea trout (*Salmo trutta*). *Molecular Ecology* **12**: 2057–2071.

Palm, S., Laikre, L. *et al.* (2003b) Effective population size and temporal genetic change in stream resident brown trout (*Salmo trutta*, L.). *Conservation Genetics* **4**: 249–264.

Palumbi, S.R. (2001) Humans as the world's greatest evolutionary force. *Science* **293**: 1786–1790.

Pamilo, P. and Nei, M. (1988) Relationships between gene trees and species trees. *Molecular Biology and Evolution* **5**: 568–583.

Panda, S., Hogenesch, J.B. *et al.* (2002) Circadian rhythms from flies to human. *Nature* **417**: 329–335.

Parmesan, C. and Yohe, G. (2003) A globally coherent fingerprint of climate change impacts across natural systems. *Nature* **421**: 37–42.

Pastor, T., Garza, J. *et al.* (2004) Low genetic variability in the highly endangered Mediterranean monk seal. *Journal of Heredity* **95**: 291–300.

Paterson, S., Wilson, K. *et al.* (1998) Major histocompatibility complex variation associated with juvenile survival and parasite resistance in a large unmanaged ungulate population (*Ovis aries* L.). *Evolution* **95**: 3714–3719.

Peakall, R. and Smouse, P.E. (2006) GENALEX 6: genetic analysis in Excel. Population genetic software for teaching and research. *Molecular Ecology Notes* **6**: 288–295.

Peichel, C.L., Ross, J.A. *et al.* (2004) The master sex-determination locus in threespine sticklebacks is on a nascent Y chromosome. *Current Biology* **14**: 1416–1424.

Pemberton, J.M. (2004) Measuring inbreeding depression in the wild: the old ways are the best. *Trends in Ecology and Evolution* **19**: 613–615.

Pemberton, J.M. (2008) Wild pedigrees: the way forward. *Proceedings of the Royal Society of London Series B Biological Sciences* **275**: 613–621.

Pemberton, J.M., Coltman, D.W. *et al.* (1999) Using microsatellites to measure the fitness consequences of inbreeding and outbreeding. In *Microsatellites: Evolution and Applications*, D.B. Goldstein and C. Schlötterer (eds), pp. 151–162. Oxford: Oxford University Press.

Penn, D.J. (2002) The scent of genetic compatibility: sexual selection and the major histocompatibility complex. *Ethology* **108**: 1–21.

Penn, D.J. and Potts, W.K. (1998) How do major histocompatibility genes influence odor and mating preferences? *Advances in Immunology* **69**: 411–435.

Penn, D.J. and Potts, W.K. (1999) The evolution of mating preferences and major histocompatibility complex genes. *American Naturalist* **153**: 145–164.

Pertoldi, C., Bijlsma, R. *et al.* (2007) Conservation genetics in a globally changing environment: present problems, paradoxes and future challenges. *Biodiversity and Conservation* **16**: 4147–4163.

Petersson, E. and Järvi, T. (1993) Differences in reproductive traits between sea-ranched and wild seatrout (*Salmo trutta*) originating from a common stock. *Nordic Journal of Freshwater Research* **68**: 91–97.

Petersson, E. and Järvi, T. (1995) Evolution of morphological traits in sea trout (*Salmo trutta*) parr (0+) through sea-ranching. *Nordic Journal of Freshwater Research* **70**: 62–67.

Petersson, E., Järvi, T. *et al.* (1996) The effect of domestication on some life history traits of sea trout and Atlantic salmon. *Journal of Fish Biology* **48**: 776–791.

Petit, C., Fréville, H. *et al.* (2001) Gene flow and local adaptation in two endemic plant species. *Biological Conservation* **100**: 21–34.

Phillips, B., Brown, G. *et al.* (2003) Assessing the potential impact of cane toads on Australian snakes. *Conservation Biology* **17**: 1738–1747.

Phillips, B., Brown, G. *et al.* (2006a) Invasion and the evolution of speed in toads. *Nature* **439**: 803.

Phillips, B. and Shine, R. (2006b) An invasive species induces rapid adaptive change in a native predator: cane toads and black snakes in Australia. *Proceedings of the Royal Society of London Series B Biological Sciences* **273**: 1545–1550.

Phillips, B.L. and Shine, R. (2004) Adapting to an invasive species: toxic cane toads induce morphological change in Australian snakes. *Proceedings of the National Academy of Sciences USA* **101**: 17150–17155.

Piertney, S., Maccoll, A. *et al.* (1998) Local genetic structure in red grouse (*Lagopus lagopus scoticus*): evidence from microsatellite DNA markers. *Molecular Ecology* **7**: 1645–1654.

Piertney, S.B. and Oliver, M.K. (2006) The evolutionary ecology of the major histocompatibility complex. *Heredity* **96**: 7–21.

Piertney, S.B. and Webster, L.M. (2008) Characterising functionally important and ecologically meaningful genetic diversity using a candidate gene approach. *Genetica* (in press).

Pimm, S.L. (1991) *The Balance of Nature?: Ecological Issues in Conservation of Species and Communities*. Chicago, IL: University of Chicago Press.

Pimm, S.L., Dollar, L. *et al.* (2006a) The genetic rescue of the Florida panther. *Animal Conservation* **9**: 115–122.

Pimm, S.L., Bass, O.L. *et al.* (2006b) Ockham and Garp. reply to Maehr et al's (2006) response to Pimm *et al.* (2006) *Animal Conservation* **9**: 133–134.

Piry, S., Luikart, G. *et al.* (1999) BOTTLENECK: a computer program for detecting recent reductions in the effective population size using allele frequency data. *Journal of Heredity* **90**: 502–503.

Pounds, A.J., Bustamante, M.R. et al. (2006) Widespread amphibian extinctions from epidemic disease driven by global warming. *Nature* **439**: 161–167.

Pritchard, J.K., Stephens, M. et al. (2000) Inference of population structure using multilocus genotype data. *Genetics* **155**: 945–959.

Quinn, G. and Keough, M. (2002) *Experimental Design and Data Analysis for Biologists*. Cambridge: Cambridge University Press.

Quinn, T. (2001) Evolution of chinook salmon (*Oncorhynchus tshawytscha*) populations in New Zealand: pattern rate, and process. *Genetica* **112–113**: 493–513.

Rannala, B. and Mountain, J. (1997) Detecting immigration by using multilocus genotypes. *Proceedings of the National Academy of Sciences USA* **94**: 9197–9201.

Rannala, B. and Z. Yang (2003) Bayes estimation of species divergence times and ancestral population sizes using DNA sequences from multiple loci. *Genetics* **164**: 1645–1656.

Räsänen, K., Laurila, A. et al. (2003a) Geographic variation in acid stress tolerance of the moor frog, *Rana arvalis*. I.Local adaptation. *Evolution* **57**: 352–362.

Räsänen, K., A. Laurila, et al. (2003b) Geographic variation in acid stress tolerance of the moor frog, *Rana arvalis*. II. Adaptive maternal effects. *Evolution* **57**: 363–371.

Raup, D.M. (1992) *Extinction: Bad Genes or Bad Luck?* London: W.W. Norton.

Raup, D.M. (1994) The role of extinction in evolution. *Proceedings of the National Academy of Sciences USA* **91**: 6758–6763.

Reed, D.H. and Bryant, E.H. (2000) Experimental tests of minimum viable population size. *Animal Conservation* **3**: 7–14.

Reed, D.H. and Frankham, R. (2001) Correlation between fitness and genetic diversity. *Conservation Biology* **17**: 230–237.

Reed, D.H., Briscoe, D.A. et al. (2002) Inbreeding and extinction: the effect of environmental stress and lineage. *Conservation Genetics* **3**: 301–307.

Reed, D.H., Lowe, E.H. et al. (2003) Inbreeding and extinction: effects of rate of inbreeding. *Conservation Genetics* **4**: 405–410.

Reith, W. and Mach, B. (2001) The bare lymphocyte syndrome and the regulation of MHC expression. *Annual Review of Immunology* **19**: 331–373.

Reusch, T., Häberli, M. et al. (2001) Female sticklebacks count alleles in a strategy of sexual selection explaining MHC polymorphism. *Nature* **414**: 300–302.

Reznick, D. (1997) Evaluation of the rate of evolution in natural populations of guppies (*Poecilia reticulata*). *Science* **275**: 1934–1937.

Reznick, D. and Travis, J. (1996) The empirical study of adaptation in natural populations. *Adaptation*, M.R. Rose and G.V. Lauder (eds), pp. 243–290. New York: Academic Press.

Reznick, D. and Ghalambor, C. (2001) The population ecology of contemporary adaptations: what empirical studies reveal about the conditions that promote rapid evolution. *Genetica* **112**: 183–198.

Reznick, D., Butler Iv, M.J. et al. (2001) Life-history evolution in guppies. VII. The comparative ecology of high- and low-predation environments. *American Naturalist* **157**: 126–140.

Reznick, D.N. and Bryga, H. (1987) Life-history evolution in guppies (*Poecilia reticulata*): 1. Phenotypic and genetic changes in an introduction experiment. *Evolution* **41**: 1370–1385.

Richardson, D.S. and Westerdahl, H. (2003) MHC diversity in two *Acrocephalus* species: the outbred great reed warbler and the inbred Seychelles warbler. *Molecular Ecology* **12**: 3523–3529.

Richman, A.D., Herrera, G. et al. (2007) Evidence for balancing selection at the DAB locus in the axolotl, *Ambystoma mexicanum*. *International Journal of Immunogenetics* **34**: 475–478.

Rieseberg, L. and Buerkle, C. (2002) Genetic mapping in hybrid zones. *American Naturalist* **159**: 36–50.

Riley, S., Pollinger, J. et al. (2006) A southern California freeway is a physical and social barrier to gene flow in carnivores. *Molecular Ecology* **15**: 1733–1741.

Robinson, J., Waller, M.J. et al. (2003) MGT/HLA and IMGT/MHC: sequence databases for the study of the major histocompatibility complex. *Nucleic Acids Research* **31**: 311–314.

Rogell, B. (2005) *Microsatellite Variation in the Natterjack Toad on the Swedish West-coast*. MSc thesis, Uppsala University, Uppsala.

Root, T., Price, J. et al. (2003) Fingerprints of global warming on wild animals and plants. *Nature* **421**: 57–60.

Rose, M.R. and Lauder, G.V. (eds) (1996) *Adaptation*. New York: Academic Press.

Rossiter, S.J., Jones, G. et al. (2001) Outbreeding increases offspring survival in wild greater horseshoe bats (*Rhinolophus ferrumequinum*). *Proceedings of the Royal Society of London Series B Biological Sciences* **268**: 1055–1061.

Rothermel, B. and Semlitsch, R. (2002) An experimental investigation of landscape resistance of forest versus old-field habitats to emigrating juvenile amphibians. *Conservation Biology* **16**: 1324–1332.

Roux, F., Touzet, P. et al. (2006) How to be early flowering: an evolutionary perspective. *Trends in Plant Science* **11**: 375–381.

Rubridge, E., Corbett, P. et al. (2001) A molecular analysis of hybridization between native westslope cutthroat trout and introduced rainbow trout in southeastern British Columbia. *Journal of Fish Biology* **59**: 42–54.

Rudd, S. (2003) Expressed sequence tags: alternative or complement to whole genome sequences? *Trends in Plant Science* **8**: 321–329.

Ruijter, J., Baas, A. et al. (2002) Statistical evaluation of SAGE libraries: consequences for experimental design. *Physiological Genomics* **29**: 37–44.

Ryder, O. (1986) Species conservation and systematics: the dilemma of subspecies. *Trends in Ecology and Evolution* **1**: 9–10.

Saccheri, I., Kuussaari, M. et al. (1998) Inbreeding and extinction in a butterfly metapopulation. *Nature* **392**: 491–494.

Saether, S.A., Fiske, P. et al. (2007) Inferring local adaptation from Q_{ST}-F_{ST} comparisons: neutral genetic and quantitative trait variation in European populations of Great Snipe. *Journal of Evolutionary Biology* **20**: 1563–1576.

Sahlsten, J., Thörngren, H. et al. (2008) Inference of hazel grouse populations structure using multilocus data: a landscape genetic approach. *Heredity* (in press).

Sammut, B., Laurens, V. et al. (1997) Isolation of MHC class I cDNAs from the axolotl *Ambystoma mexicanum*. *Immunogenetics* **45**: 285–294.

Sammut, B., Du Pasquier, L. *et al.* (1999) Axolotl MHC architecture and polymorphism. *European Journal of Immunology* **29**: 2897–2907.

Scheiner, S.M. (1993) Genetics and evolution of phenotypic plasticity. *Annual Review of Ecology and Systematics* **24**: 35–68.

Schemske, D.W. and Lande, R. (1987) On the evolution of plant mating systems: a reply to Waller. *American Naturalist* **130**: 804–806.

Schierup, M.H. (1998) The number of self-incompatibility alleles in a finite, subdivided population. *Genetics* **149**: 1153–1162.

Schwartz, M., Luikart, G. *et al.* (2006) Genetic monitoring as a promising tool for conservation and management. *Trends in Ecology and Evolution* **22**: 25–33.

Seddon, J. and Baverstock, P. (1999) Variation on islands: major histocombatibility complex (Mhc) polymorphism in population of the Australian bush rat. *Molecular Ecology* **8**: 2071–2079.

Segelbacher, G. and Storch, I. (2002) Capercaillie in the Alps: genetic evidence of metapopulation structure and population decline. *Molecular Ecology* **11**: 1669–1677.

Segelbacher, G. and Höglund, J. (2008) Ecological genomics and conservation – where do we stand? *Genetica* (in press).

Segelbacher, G., Höglund, J. *et al.* (2003) From connectivity to isolation: genetic consequences of population fragmentation in capercaillie across Europe. *Molecular Ecology* **12**: 1773–1780.

Segelbacher, G., Manel, S. *et al.* (2008) Temporal and spatial analyses disclose consequences of habitat fragmentation on the genetic diversity in capercaillie (*Tetrao urogallus*). *Molecular Ecology* **17**: 2356–2367.

Shapiro, M., Marks, M. *et al.* (2004) Genetic and developmental basis of evolutionary pelvic reduction in threespine sticklebacks. *Nature* **428**: 717–723.

Shiina, T., Hosomichi, K. *et al.* (2006) Comparative genomics of the poultry major histocompatibility complex. *Animal Science Journal* **77**: 151–162.

Shindo, C., Aranzana, M.J. *et al.* (2005) Role of FRIGIDA and FLOWERING LOCUS C in determining variation in flowering time of *Arabidopsis*. *Plant Physiology* **138**: 1163–1173.

Simberloff, D. (1998) Flagships, umbrellas, and keystones: is single-species management passe in the landscape era? *Biological Conservation* **83**: 247–257.

Slade, R.W. and McCallum, H.I. (1992) Overdominant vs. frequency-dependent selection at MHC loci. *Genetics* **132**: 861–862.

Slate, J. (2005) Quantitative trait locus mapping in natural populations: progress, caveats and future directions. *Molecular Ecology* **2**: 363–379.

Slate, J. and Pemberton, J.M. (2002) Comparing molecular measures for detecting inbreeding depression. *Journal of Evolutionary Biology* **15**: 20–31.

Slate, J., Kruuk, L.E.B. *et al.* (2000) Inbreeding depression influences lifetime breeding success in a wild population of red deer (*Cervus elaphus*). *Proceedings of the Royal Society of London Series B Biological Sciences* **267**: 1657–1662.

Slate, J., David, P. *et al.* (2004) Understanding the relationship between the inbreeding coefficient and multilocus heterozygosity: theoretical expectations and empirical data. *Heredity* **93**: 255–265.

Slate, J., Gratten, J. *et al.* (2008) Gene mapping in the wild with SNPs: guidelines and future directions. *Genetica* (in press).

Slatkin, M. (1985) Gene flow in natural populations. *Annual Review of Ecology and Systematics* **16**: 393–430.

Slatkin, M. and Barton, N.H. (1989) A comparison of three indirect methods for estimating the average level of gene flow. *Evolution* **43**: 1349–1368.

Slotte, T., Holm, K. *et al.* (2007) Differential expression of genes important for adaptation in *Capsella bursa-pastoris* (Brassicaceae). *Plant Physiology* **145**: 160–73.

Smith, T. and Bernatchez, L. (2008) Evolutionary change in human-altered environments. *Molecular Ecology* **17**: 1–8.

Smith, T., Bruford, M. *et al.* (1993) The preservation of process: the missing element of conservation programs. *Biodiversity Letters* **1**: 164–167.

Smith, T.B. and L. Bernatchez (2008) Evolutionary change in human-altered environments. *Molecular Ecology* **17**: 1–8.

Smith, T.B., Wayne, R.K. *et al.* (1997) A role for ecotones in generating rainforest biodiversity. *Science* **276**: 1855–1857.

Söderhäll, K. (2004) Krasslig kräfta, tvivelaktig import. *Miljöforskning*. Stockholm: The Swedish Research Council for Environment, Agricultural Sciences and Spatial Planning.

Sodhi, N., Liow, L. *et al.* (2004) Avian extinction from tropical and subtropical forests. *Annual Review of Ecology, Evolution, and Systematics* **35**: 323–345.

Sommer, S. (2005) The importance of immune gene variability (MHC) in evolutionary ecology and conservation. *Frontiers in Zoology* **2**: 16.

Soulé, M. (1976) Allozyme variation: its determinant in space and time. In *Molecular Evolution*, F.J. Ayala (ed.), pp. 60–76. Sunderland, MA: Sinauer Associates.

Soulé, M.E. (1980) Thresholds for survival: maintaining fitness and evolutionary potential. *Conservation Biology: an Evolutionary-Ecological Perspective*. M.E. Soulé and B.A. Wilcox, pp. 151–169. Sunderland, MA: Sinauer Associates.

Soulé, M.E. (ed.) (1986) *Conservation Biology: the Science of Scarcity and Diversity*. Sunderland, MA: Sinauer Associates.

Soulé, M.E. (1987) *Viable Populations for Conservation*. Cambridge: Cambridge University Press.

Soulé, M.E. and Mills, L.S. (1998) Population genetics: no need to isolate genetics. *Science* **282**: 1658–1659.

Spear, S., Peterson, C. *et al.* (2005) Landscape genetics of the blotched tiger salamander (*Ambystoma tigrinum melanostictum*). *Molecular Ecology* **14**: 2553–2564.

Spear, S., Peterson, C. *et al.* (2006) Molecular evidence for historical and recent population size reductions of tiger salamanders (*Ambystoma tigrinum*) in Yellowstone National Park. *Conservation Genetics* **7**: 605–611.

Spielman, D., Brook, B.W. *et al.* (2004) Most species are not driven to extinction before genetic factors impact them. *Proceedings of the National Academy of Sciences USA* **101**: 15261–15264.

Spitze, K. (1993) Population structure in *Daphnia obtusa*: quantitative genetic and allozymic variation. *Genetics* **135**: 367–374.

Stearns, S.C. and Koella, J.C. (1986) The evolution of phenotypic plasticity in life-history traits: predictions of reaction norms for age and size at maturity. *Evolution* **40**: 893–913.

Stet, R., Kruiswijk, C. *et al.* (2003) Major histocompatibility lineages and immune gene function in teleost fishes: the road not taken. *CRC Critical Reviews in Immunology* **23**: 441–471.

Stinchcombe, J.R., Weinig, C. *et al.* (2004) A latitudinal cline in flowering time in *Arabidopsis thaliana* modulated by the flowering time gene FRIGIDA. *Proceedings of the National Academy of Sciences USA* **101**: 4712–4717.

Stockwell, C.A., Hendry, A.P. *et al.* (2003) Contemporary evolution meets conservation biology. *Trends in Ecology and Evolution* **18**: 294–301.

Storfer, A., Murphy, M. *et al.* (2007) Putting the 'landscape' in landscape genetics. *Heredity* **98**: 128–142.

Storz, J.F. (2002) Contrasting patterns of divergence in quantitative traits and neutral DNA markers: analysis of clinal variation. *Molecular Ecology* **11**: 2537–2551.

Storz, J.F. (2005) Using genomic scans of DNA polymorphism to infer adaptive population divergence. *Molecular Ecology* **14**: 671–688.

Strand, T., Westerdahl, H. *et al.* (2007) The Mhc class II of the Black grouse (*Tetrao tetrix*) consists of low numbers of B and Y genes with variable diversity and expression. *Immunogenetics* **59**: 725–734.

Suzuki, Y. (2004) New methods for detecting positive selection at single amino acid sites. *Journal of Molecular Evolution* **59**: 11–19.

Swaddle, J. and Lockwood, R. (1998) Morphological adaptations to predation risk in passerines. *Journal of Avian Biology* **29**: 172–176.

Szarowska, M., Falniowski, A. *et al.* (1998) Adaptive significance of glucose phosphate isomerase (GPI) allozymes in the spring snail *Bythinella*? *Journal of Molluscan Studies* **64**: 257–261.

Taberlet, P., Swenson, J. *et al.* (1995) Localization of a contact zone between two highly divergent mitochondrial DNA lineages of the brown bear (*Ursus arctos*) in Scandinavia. *Conservation Biology* **9**: 1255–1264.

Tajima, F. (1989) Statistical-method for testing the neutral mutation hypothesis by DNA polymorphism. *Genetics* **123**: 585–595.

Takahata, N. and Nei, M. (1990) Allelic geneology under overdominant and frequency-dependent selction and polymorphism of major histocompatibility complex loci. *Genetics* **124**: 967–978.

Taylor, E.B., Stamford, M.D. *et al.* (2003) Population subdivision in westslope cutthroat trout (*Oncorhynchus clarki lewisi*) at the northern periphery of its range: evolutionary inferences and conservation implications. *Molecular Ecology* **12**: 2609–2622.

Thomas, C., Cameron, A. *et al.* (2004) Extinction risk from climate change. *Nature* **427**: 145–148.

Thomas, J., Telfer, M. *et al.* (2004) Comparative losses of british butterflies, birds, and plants and the global extinction crisis. *Science* **303**: 1879–1881.

Thompson, J.N. (1996) Evolutionary ecology and the conservation of biodiversity. *Trends in Ecology and Evolution* **11**: 300–303.

Thompson, K. and Jones, A. (1999) Human population density and prediction of local plant extinction in Britain. *Conservation Biology* **13**: 185–189.

Thörngren, H. (2006) *Genetic Structure in Northern Fringe Populations of the Natterjack Toad, Bufo calamita*. MSc thesis, Uppsala University.

Townsend-Peterson, A., Ortega-Huerta, M. *et al.* (2002) Future projections for Mexican faunas under global climate change scenarios. *Nature* **416**: 626–629.

Travers, S.E., Smith, M.D. *et al.* (2007) Ecological genomics: making the leap from model systems in the lab to native populations in the field. *Frontiers in Ecology and the Environment* **5**: 19–24.

Tregenza, T. and Wedell, N. (2000) Genetic compatibility, mate choice and patterns of parentage. *Molecular Ecology* **9**: 1013–1027.

Tsutsui, N. and Case, T. (2001) Population genetics and colony structure of the Argentine ant (*Linepithema humile*) in its native and introduced ranges. *Evolution* **55**: 976–985.

Ujvari, B., Madsen, T. *et al.* (2002) Low genetic diversity threatens imminent extinction for the Hungarian meadow viper (*Vipera ursinii rakosiensis*). *Biological Conservation* **105**: 127–130.

Urban, M.C., Phillips, B.L. *et al.* (2008) A toad more traveled: the heterogeneous invasion dynamics of cane toads in Australia. *American Naturalist* **171**: E134–E148.

Våge, D.I., Fuglei, E. *et al.* (2005) Two cysteine substitutions in the MC1R generate the blue variant of the arctic fox (*Alopex lagopus*) and prevent expression of the white winter coat. *Peptides* **26**: 1814–1817.

Van Dongen, S. *et al.* (1997) Genetic population structure of the winter moth (*Operophtera brumata* L.) (Lepidoptera, Geometridae) in a fragmented landscape. *Heredity* **80**: 92–100.

van Noordwijk, A.J. and Scharloo, W. (1981) Inbreeding in an island population of the great tit. *Evolution* **35**: 674–688.

van Oosterhout, C., Joyce, D.A. *et al.* (2006) Balancing selection, random genetic drift, and genetic variation the major histocompatability complex in two wild populations of guppies (*Poecilia reticulata*). *Evolution* **60**: 2562–2574.

van Oosterhout, C., Smith, A. *et al.* (2007) The guppy as a conservation model: implications of parasitism and inbreeding on reintroduction success. *Conservation Biology* **21**: 1573–1583.

van Tienderen, P.H., de Haan, A.A. *et al.* (2002) Biodiversity assessment using markers for ecologically important traits. *Trends in Ecology and Evolution* **17**.

van Tienderen, P.H. and van der Toorn, J. (1991) Genetic differentiation between populations of *Plantago lanceolata*. I. Local adaptation in three contrasting habitats. *Journal of Ecology* **79**: 27–42.

Vasemägi, A. and Primmer, C. (2005) Challenges for identifying functionally important genetic variation: the promise of combining complementary research strategies. *Molecular Ecology* **14**: 3623–3642.

Vekemans, X. and Slatkin, M. (1994) Gene and allelic genealogies at a gametophytic self-incompatibility locus. *Genetics* **137**: 1157–1165.

Vera, J., Wheat, C. *et al.* (2008) Rapid transcriptome characterization for a nonmodel organism using 454 pyrosequencing. *Molecular Ecology* **17**: 1636–1647.

Via, S. and Lande, R. (1985) Genotype-environment interaction and the evolution of phenotypic plasticity. *Evolution* **39**: 505–522.

Vilà, C., Sundqvist, A.-K. *et al.* (2003) Rescue of a severely bottlenecked wolf (*Canis lupus*) population by a single immigrant. *Proceedings of the Royal Society of London Series B Biological Sciences* **270**: 91–97.

Vincek, V., Klein, D. *et al.* (1995) Molecular-cloning of major histocompatibility complex class-II B-gene cDNA from the Bengalese finch *Lonchura striata*. *Immunogenetics* **42**: 262–267.

Vogler, A.P. and DeSalle, R. 1994. Diagnosing units of conservation management. *Conservation Biology* **6**: 170–178.

Vos, P., Hogers, R. *et al.* (1995) AFLP: an new technique for DNA fingerprinting. *Nucleic Acids Research* **23**: 4407–4414.

Wang, M. and Schreiber, A. (2001) The impact of habitat fragmentation and social structure on the population genetics of roe deer (*Capreolus capreolus* L.) in central Europe. *Heredity* **86**: 703–715.

Wang, Y., Williams, D. *et al.* (2005) Evidence for a recent genetic bottleneck in the endangered Florida Keys silver rice rat (*Oryzomys argentatus*) revealed by microsatellite DNA analyses. *Conservation Genetics* **6**: 575–585.

Waples, R.S. 1991. Pacific salmon, *Oncorhynchus* spp. and the definition of "species" under the endangered species act. *Marine Fisheries Reviews* **53**: 11–22.

Waples, R.S. and Gaggiotti, O.E. (2006) What is a population? An empirical evaluation of some genetic methods for identifying the number of gene pools and their degree of connectivity. *Molecular Ecology* **15**: 1419–1439.

Waples, R.S., Zabel, R.W. *et al.* (2008) Evolutionary responses by native species to major anthropogenic changes to their ecosystems: Pacific salmon in the Columbia River hydropower system. *Molecular Ecology* **17**: 84–96.

Watt, W. and Dean, A. (2000) Molecular-fuctional studies of adaptive genetic variation in procaryotes and eucaryotes. *Annual Reviewsin Genetics* **34**: 593–622.

Watterson, G. (1977) Heterosis or neutrality. *Genetics* **85**: 789–814.

Watterson, G. (1984) Allele frequencies after a bottleneck. *Theoretical Population Biology* **26**: 387–407.

Weber, A., Weber, K. *et al.* (2007) Sampling the *Arabidopsis* transcriptome with massively parallel pyrosequencing. *Plant Physiology* **144**: 32–42.

Weber, D., Stewart, B. *et al.* (2004) Major histocompatibility complex variation at three class II loci in the northern elephant seal. *Molecular Ecology* **13**: 711–718.

Wedekind, C. and Füri, S. (1997) Body odour preferences in men and women: do they aim for specific MHC combinations or simply heterozygosity? *Proceedings of the Royal Society of London Series B Biological Sciences* **264**: 1387–1387.

Weir, B.S. and Cockerham, C. (1984) Estimating F-statistics for the analysis of population structure. *Evolution* **38**: 1358–1370.

Westerdahl, H., Wittzell, H. *et al.* (1999) Polymorphism and transcription of Mhc class I genes in a passerine bird, the great reed warbler. *Immunogenetics* **49**: 158–170.

Westerdahl, H., Wittzell, H. *et al.* (2000) Mhc diversity in two passerine birds: no evidence far a minimal essential Mhc. *Immunogenetics* **52**: 92–100.

Westmeier, R.L., Brawn, J.D. *et al.* (1998) Tracking the long-term decline and recovery of an isolated population. *Science* **282**: 1695–1698.

Wheat, C. (2008) Rapidly developing functional genomics in ecological model systems via 454 transcriptome sequencing. *Genetica* (in press).

Whitehead, A. and Crawford, D. (2007) Variation within and among species in gene expression: raw material for evolution. *Molecular Ecology* **15**: 1197–1211.

Wilding, C.S., Butlin, R.K. *et al.* (2001) Differential gene exchange between parapatric morphs of *Littorina saxatilis* detected using AFLP markers. *Journal of Evolutionary Biology* **14**: 611–619.

Willi, Y., Van Buskirk, J. *et al.* (2006) Limits to the adaptive potential of small populations. *Annual Review of Ecology, Evolution and Systematics* **37**: 433–458.

Wilson, E.O. (1992) *The Diversity of Life*. Cambridge, MA: Harvard University Press.

Wirth, T. and Bernatchez, L. (2001) Genetic evidence against panmixia in the European eel. *Nature* **409**: 1037–1040.

Withler, R., Beacham, T. *et al.* (1997) Species identification of Pacific salmon by PCR-RFLP of a MHC gene. *North American Journal of Fisheries and Managent* **17**: 929–938.

Wittzell, H., von Schantz, T. *et al.* (1995) Rfp-Y-like sequences assort independently of pheasant Mhc genes. *Immunogenetics* **42**: 68–71.

Wofford, J., Gresswell, R. *et al.* (2005) Influence of barriers to movement on within-watershed genetic variation of coastal cutthroat trout. *Ecological Applications* **15**: 628–637.

Wren, J., Forgacs, E. *et al.* (2000) Repeat polymorphisms within gene regions: phenotypic and evolutionary implications. *American Journal of Human Genetics* **67**: 345–356.

Wright, D., Kerje, S. *et al.* (2008) The genetic architecture of a female sexual ornament. *Evolution* **62**: 86–98.

Wright, S. (1921) Correlation and causation. *Journal of Agricultural Research* **20**: 557–585.

Wright, S. (1922) Coefficients of inbreeding and relationship. *American Naturalist* **56**: 330–338.

Wright, S. (1929) Fisher's theory of dominance. *American Naturalist* **63**: 274–279.

Wright, S. (1931) Evolution in mendelian populations. *Genetics* **16**: 97–159.

Wright, S. (1938) Size of population and breeding structure in relation to evolution. *Science* **87**: 430–431.

Wright, S. (1951) The genetical structure of populations. *Annals of Eugenics* **15**: 323–354.

Wright, S. (1969) *The Theory of Gene Frequencies*. Chicago, IL: University of Chicago Press.

Wu, P., T.-X. Jiang, *et al.* (2004) Molecular shaping of the beak. *Science* **305**: 1465–1466.

Yan, H. and Zhou, W. (2004) Allelic variations in gene expression. *Current Opinion in Oncology* **16**: 39–43.

Yang, Z. and Nielsen, R. (2000) Estimating synonymous and nonsynonymous substitution rates under realistic evolutionary models. *Molecular Biology and Evolution* **17**: 32–43.

Yang, Z., Nielsen, R. *et al.* (2000) Codon substitution models for heterogeneous selection pressures at amino acid sites. *Genetics* **155**: 431–449.

Young, T.P. (1991) Diversity overrated. *Nature* **352**: 10.

Zelano, B. and Edwards, S.V. (2002) An Mhc component to kin recognition and mate choice in birds: predictions, progress, and prospects. *American Naturalist* **160**: S225–S237.

Ziegler, A., Kentenich, H. *et al.* (2005) Female choice and the MHC. *Trends in Immunology* **26**: 496–502.

Zoorob, R., Bernot, A. *et al.* (1993) Chicken major histocompatibility complex class-II B-genes—analysis of interallelic and interlocus sequence variance. *European Journal of Immunology* **23**: 1139–1145.

Index

Note: all species are indexed on both common and Latin names.

acid environments 105, 107
Aconyx jubatus 85
Acrocephalus 86
adaptation, local 102–18
adaptive evolutionary conservation 147
adder 89
admixture mapping 126
African cheetah 85
Agelaius phoeniceus 86
Agrostis tenuis 104
alien species 140
alleles
  null 40, 42
  per locus 22
allele-specific analysis 126
allelic richness 22
allozyme variation 18–19
Alpine newt 91
Ambystoma
  A. mexicana 88
  A. tigrinum 88
  A. tigrinum melanostictum 71, 72, 74
American crayfish 14
amphibians, *Mhc* genes and conservation 88–91
amplified fragment length polymorphism 20–1, 24, 125, 131
Anguilla anguilla 65
annotation 121–2
Anolis lizard 104
Aphanomyces astaci 14
Arabidopsis thaliana 98–9, 122, 124
Arabis petraea 11
Argentine ant 80
Arnica montana 11
aspen 100
assembly 121–2
association analysis 127
assortative mating 38
Atlantic salmon 92
Attwater's prairie chicken 7
Australian black snake 80
axolotl 88

banner-tailed kangaroo rat 75
Barbary red deer 74
Bayesian inference 69
Bengalese finch 86
bighorn sheep 67
biodiversity 1, 139
birds, *Mhc* genes and conservation 85–8
Biston betularia 96, 104
black grouse 44, 64, 74
  genetic diversity 45
  lekking 55
black-tailed godwit 145
blotched tiger salamander 71, 72, 74
bluethroat 97
blue tit 97, 104
Bmp4 136
bobcat 67
bobwhite quail 86
Bombina bombina 90
Bombyx mori 122
Bonasa bonasia 71, 73
bottlenecks 73–5
Brassica
  B. insularis 112, 114
  B. nigra 100
breeder's equation 31, 32
brown bear 65–6, 73
brown trout 49, 93–4
Bufo
  B. calamita 40, 43, 74, 107
  B. marinus 79
  B. viridis 58–9

California valley oak 145
calmodulin 138
candidate genes 123
cane toad 79
Canis
  C. latrans 67
  C. lupus 8
capercaillie 13–14, 50, 61, 64, 73
Capreolus capreolus 67, 70
Castor fiber 12
Centaurea corymbosa 112, 114

Cepaea snails 95–6, 104
Cervus
  C. elaphus 12, 53
  C. elaphus barbarus 74
Chatham Island black robin 87
chinook salmon 92
Clarkia pulchella 15
Clock gene 97–100
codominant neutral variation 18–23
Colinus virginianus 86
collared flycatcher 52
colonizing species 77
common frog 105, 107, 114, 134
common shrew 67
conservation
  Mhc genes
    birds 85–8
    fish 91–4
    mammals 84–5
    reptiles and amphibians 88–91
  quantitative trait differentiation 114–16
conservation units 145–8
contigs 121
core anchor tagged sequences 130
correspondence analysis 62
Coturnix japonica 97
coyote 67
Cyanistes caeruleus 97, 104
Cynopterus sphinx 115

Danio rerio 91
Darwin's finch 104, 136–8
diapause 105
differential gene expression 132–3
Dipodomys spectabilis 75
direct-effect hypothesis 57
DNA slippage 129
domestic chicken 122
dominant neutral markers 23–4
Drosophila melanogaster 15–16, 35, 104, 132–3

ecological genomics 119–38
  assembly and annotation 121–2
  evolutionary and ecological analyses 122–9
  whole-genome sequencing 120–1
ectodysplasin A 135
eel 65
effective population size 47–9
El Niño events 87
endangered species 12, 147
  population structure 50
  see also extinction
Endangered Species Act 147
environmental association analysis 126
eumelanin 96
Eurasian beaver 85
evening primrose 15
evolution 105

human impact on 140–2
  rates of 140, 141
evolutionarily significant units 145, 146
evolutionary potential 143–5
Ewens-Watterson test 124
exon priming intron crossing 130
expressed sequence tags 120, 121–2
extinction 1, 139
  causes of 7–8
  experimental studies 14–16
  and genetic variation 5–14
extinction vortex 2–5, 7

Falco punctatus 12
Ficedula albicollis 52
fire-bellied toad 90
fish
  harvest-induced changes 142–3
  hatchery rearing 142
  Mhc genes and conservation 91–4
Fisher's theorem 32, 34
fitness, and heterozygosity 55–7
flounder 140–1
frequency-dependent selection 83
fruit fly 15–16, 35, 104, 132–3

Galápagos penguin 87
Gallinago media 26, 86, 88, 115
Gallus gallus 122
Gammarus duebeni 16
Gasterosteus aculeatus 91, 134–5
genes
  candidate 123
  housekeeping 122
  Mhc 82–95
  see also individual genes
gene expression, differential 132–3
gene flow 23
  human-induced barriers to 60–9
general/global-effect hypothesis 56–7
genetic diversity 60–80
genetic drift 3, 4, 42, 50
genetic load 51
genetic markers 20–1
genetic variation 1
  experimental studies 14–16
  and extinction 5–14
  loss of 3–4
  measurement of 18–36
    codominant neutral variation 18–23
    dominant neutral markers 23–4
    non-neutral markers and neutrality tests 26–7
    sequence variation 24–6
  plants 6
  quantitative additive 27–36
genomics
  in conservation 129–34
  ecological 119–38

non-model species 134–8
population 127, 128
Gentiana pneumonanthe 11
geographical information systems 70
Geospiza 104, 136–8
Glanville fritillary 12, 13, 121
grayling 112, 113
greater prairie chicken 9, 10
great snipe 26, 86, 88, 115
great tit 52
green toad 58–9
Gulo gulo 66–7
guppy 91, 104
Gypsophila fastigiata 50, 105

habitat
　fragmentation 140
　loss of 60–1, 140
haldane units 140, 141
haplotype diversity 26
Hardy-Weinberg equilibrium 18, 38, 124
harvesting, evolutionary responses 142–3
Hawaiian honeycreeper 86, 87
hazel grouse 71, 73
heath hen 5–6
Heliconis erato 122
heritability 27–36
heterosis 51
heterozygosity 2, 5
　expected 22
　and inbreeding 40
　observed 22
heterozygosity-fitness correlations 55–7
heterozygote advantage 51, 83
housefly 14–15
housekeeping genes 122
house sparrow 86
Hudson-Kreitman-Aguadé test 27, 124
hunting 143

ideal population 37
immunogenetics 94–5
inbreeding 2–3, 37–44
　experimental studies 14–16
　and heterozygosity 40
inbreeding coefficient 2, 22, 40
inbreeding depression 8, 51–5
Indian fruit bat 115
industrial melanism 96, 104
intertidal snail 134
invasive species 78–80, 140
island populations 52–3

Japanese macaques 74
Japanese quail 97
junk DNA 120

Karner blue butterfly 147
Kurt's *f* value 41–2

Lacerta agilis 89
Lagopus
　L. lagopus 70
　L. lagopus scoticus 70
Lande, Russell 2
landscape genetics 69–73
leopard's bane 11
Limosa limosa limosa 145
Linephithema humile 80
linkage disequilibrium 128
Littorina saxatilis 134
local adaptation 102–18
　allochronic method 103–4
　evidence of 103–8
　synchronic method 104
local-effect hypothesis 57
Lonchura striata 86
Lotus scoparius 105
Luscinia svecica 97
Lutra lutra 70
Lycaeides
　L. melissa melissa 147
　L. melissa samuelis 147
Lynx rufus 67

Macaca fuscata 74
McDonald-Kreitman test 27, 124–5
mammals, *Mhc* genes and conservation 84–5
management units 145
markers
　dominant neutral 23–4
　genetic 20–1
　microsatellite 130
　non-neutral 26–7
marsh gentian 11
mass extinctions 1
Mauritius kestrel 12
mc1r gene 95–8
meadow viper 90
Mediterranean monk seal 74
melanin 96
melanocytes 96
melanogenesis 96–7
Melitaea cinxia 12, 13, 121
Melospiza melodia 52
Mendelian segregation 3
Mendelian traits 29
Mesotriton alpestris 91
metapopulations 45–6
Mhc genes 82–95, 134–5
　and conservation
　　birds 85–8
　　fish 91–4
　　mammals 84–5
　　reptiles and amphibians 88–91
　and immunogenetics 94–5
　organization and size 82–3
microarrays 132–3
microevolution 140

microsatellites 129
  diversity 18–19, 91–2
  markers 130
Milissa blue butterfly 147
Mirounga angustirostrus 12, 85
Monachus monachus 74
moor frog 105, 107
mRNA 120
  allele-specific analysis 126
  linkage mapping 125
multidimensional scaling 61, 62
Musca domestica 14–15
muskoxen 53, 54
mutational meltdown 4
mutations
  deleterious 4
  silent/synonymous 25

narrow-leaf plantain 105
natterjack toad 40, 43, 74, 107
natural selection 53
neutrality tests 26–7, 124–5
  multiple-marker-based 125
  sequence-based 124–5
  single-locus 124–5
neutral markers 23–4
new rares 10–11
New Zealand robin 86
non-model species 134–8
non-neutral markers 26–7
northern elephant seal 12, 85
northern rockcress 11
Norwegian red deer 12
nucleotide diversity 25–6
null alleles 40, 42

Oncorynchus
  O. mykiss 97
  O. tshawytscha 92
Operophtera brumata 67
Oryzomys argentatus 74
otter 70
outbreeding 51
overdominance hypothesis 51
overharvesting 140
Ovibus moscatus 53, 54
Ovis canadensis 67

Pacifastacus leniusculus 14
parallel pyrosequencing 120
partial dominance hypothesis 51
Parus major 52
Passerculus sanwichensis 86
Passer domesticus 86
path analysis 41
pedigree 41
peppered moth 96, 104
Peromyscus mice 104, 118
Petroica
  P. australis australis 87
  P. traversi 87
phaeomelanin 96
phenotype 28
phenotypic plasticity 140–1
photoperiodism 97–100, 116
Phylloscopus trochilus 73
phylogenetics 133–4
phylogeny 144
Phyteuma spicatum 11
pigmentation genes 95–8
pine tree 74
Pinus taeda 74
pitcher-plant mosquito 105, 106
Pitx1 136
Plantago lanceolata 105
plants, genetic variation 6
Platichtys flesus 140–1
Poecilia reticulata 91, 104
Pogonatum dentatum 77
polymorphic loci 19
population differentiation 22–3, 108
  vs quantitative trait differentiation 109–14
population expansions 75–8
population fragmentation 60–9
population genomics 127, 128
population size
  50/500 rule 4
  effective 47–9
  minimum 4
  and relative fitness 11
population structure 44–7
  Bayesian inference 69
  endangered species 50
Populus tremula 100
prairie grouse 148
principal component analysis 61, 62
principal coordinates analysis 62
proportion of variable sites 25
Psetta maxima 65, 141
Pseudechis porphyriacus 80
puma 67
Puma concolor 67
Punnet square 38

quantitative additive genetic variation 27–36
quantitative trait differentiation 108–9
  conservation studies 114–16
  vs population differentiation 109–14
quantitative trait loci 20–1, 122, 126
  mapping 131–2
  mRNA 125
  protein expression variation 126
Quercus lobata 145

rainbow trout 97
Rana
  R. arvais 105, 107
  R. temporaria 105, 107, 114, 134

randomly amplified polymorphic DNA 20–1, 23
range shifts 75–8
Rattus rattus 75
red deer 12, 53
red grouse 70
redwing blackbird 86
relative fitness, and population size 11
reproductive success 48
reptiles, *Mhc* genes and conservation 88–91
rescue effects 58–9
restriction fragment length polymorphism 20–1, 23–4
roe deer 67, 70
Rorippa
  R. amphibia 67–8
  R. palustris 67–8
  R. sylvestris 67–8

salinity 140–1
Salmo
  S. salar 92
  S. trutta 49, 93–4
salmonids 92
  hatchery rearing 142
sand lizard 89
Sanger sequencing 120
savannah sparrow 86
selective pressure 143
sequence variation 24–6
sexual reproduction 3
shrew 73
silent/synonymous mutations 25
silkworm 122
silver rice rat 74
single nucleotide polymorphisms 18, 19, 128
  detection and genotyping 129–31
song sparrow 52
Sorex araneus 67, 73
South Island robin 87
speciation 1
Spheniscus mendiculus 87
spiked rampion 11
Swedish beaver 12

Tajima's *D* test 27, 124
taxonomy 123
Tetrao
  T. tetrix 44, 45, 55, 64, 73
  T. urogallus 13–15, 50, 61, 64, 73
three-spined stickleback 91, 134–6
Thymallus thymallus 112, 113
tiger salamander 88
traits
  heritability 29–30
  Mendelian 29
  morphological 30
transcriptomics 119
Trinidadian guppy 91
trophy hunting 143
turbot 65, 141
Tympanuchus 148
  T. cupido attwateri 7
  T. cupido cupido 5–6
  T. cupido pinnatus 9, 10

Ursus arctos 65–6, 73

Vipera
  V. berus 89
  V. ursinii 90

warbler 86
whole-genome sequencing 119, 120–1
willow grouse 127–8
willow warbler 73
winter moth 67
wolf 8
wolverine 66–7
Wyemyia smithii 105, 106

Xenopus 89
  X. laevis 89
  X. tropicalis 89

zebra finch 122
zebrafish 91

גד